Smart Coatings II

ACS SYMPOSIUM SERIES **1002**

Smart Coatings II

Theodore Provder, Editor
Polymer & Coatings Consultants, LLC

Jamil Baghdachi, Editor
Eastern Michigan University

American Chemical Society, Washington, DC

ISBN: 978-0-8412-7218-7

Copyright © 2009 American Chemical Society

Distributed by Oxford University Press

Foreword

The ACS Symposium Series was first published in 1974 to provide a mechanism for publishing symposia quickly in book form. The purpose of the series is to publish timely, comprehensive books developed from ACS sponsored symposia based on current scientific research. Occasionally, books are developed from symposia sponsored by other organizations when the topic is of keen interest to the chemistry audience.

Before agreeing to publish a book, the proposed table of contents is reviewed for appropriate and comprehensive coverage and for interest to the audience. Some papers may be excluded to better focus the book; others may be added to provide comprehensiveness. When appropriate, overview or introductory chapters are added. Drafts of chapters are peer-reviewed prior to final acceptance or rejection, and manuscripts are prepared in camera-ready format.

As a rule, only original research papers and original review papers are included in the volumes. Verbatim reproductions of previously published papers are not accepted.

ACS Books Department

Contents

Overview

Bioactive Coatings

Nanotechnology

Novel Coatings

Indexes

Preface

Over the years the coatings industry has significantly evolved technologically, being driven by environmental laws, cost/performance, and perceived customer needs. We will show how these driving forces have influenced the industry to develop new technologies and that the need for advanced technologies is currently driving the development of *Smart Coatings*.

Prior to 1966 few regulations or laws restricted the amount of volatile organic compounds (VOC) or restricted the use of toxic substances. In 1966, Rule 66 was established by the Los Angeles Air Pollution Control Department regarding VOC emissions. Rule 66 was the first of much key environmental legislation (*1*) that has driven the coatings industry to meet ever-decreasing VOC targets, primarily with innovations in coatings technology. These driving forces led the coatings industry to develop four main environmentally friendly technologies to supplant traditional high VOC solvent-based coatings: waterborne coatings, powder coatings, high-solids coatings, and radiation curable coatings.

More recently the issues of green chemistry and sustainability have significantly impacted the coatings industry (*1*) driven by (1) federal government initiatives such as the passage of the Green Chemistry Research and Development Act of 2005 (HR 1215) by the U.S. House of Representatives as well as many such similar state initiatives; and (2) organizations defining the principals of Green Chemistry as well as establishing standards and certifications that a consumer product is green. Industrial input into this Green Chemistry movement has been the movement toward sustainable chemistry, which means products and processes should not only be "green" but be recyclable and reusable.

All these driving forces significantly impacted and challenged the coatings industry. Traditionally, coatings had the primary functions of protecting and decorating the substrate. More recently there has been a significant growth in the research and development of coatings and associated product development wherein the coatings have novel functions and sense and interact with their environment in addition to having the traditional protective and decorative functions. These types of coatings are often referred to as *Smart Coatings,* which can provide significant added value by optimizing cost/performance and satisfying perceived customer needs while still satisfying the requirements of environmental regulations and green sustainable chemistry.

A recently published study (*2*) in Europe looked at defining a research agenda in surface technology based upon future perceived research demands in the broad field of industrial coatings. It was found that coatings, which are multifunctional and combine several functions and/or are composed of multiple layers in order to realize property profiles with increasing complexity, were desired. For coatings, in this study, the most important desired nontraditional functions were defined as follows:

- Easy to clean and self-cleaning properties
- Scratch and mar resistance and generally adjustable tribological properties
- Coatings that enable the recycling of the coated substrate such as plastics or that can be delaminated by an external stimulus
- Switchable color, electrochromy, and electroluminescence
- Integrated product identity such as protection of trademarks or sensors for locker systems
- Antenna functions
- Heatable layers
- UV/IR absorption properties
- Biological activity such as antifouling and antibacterial
- Photovoltaic activity

Some of the enabling technologies envisioned to facilitate the achievement of the above properties include nanotechnology, encapsulation techniques, and sol–gel technology.

The properties described above, along with the associated enabling technologies just cited, are included under the rubric of *Smart Coatings*. The first *Smart Coatings* symposium in the United States was held in 2005 and the papers from that symposium are embodied in the first American Chemical Society (ACS) Symposium Series book on this subject (*3*).

Smart Coatings II begins with a current overview of this growing and active field. The first major section deals with bioactive coatings that are antibacterial and antifouling, which are achieved with a range of chemistries and morphologies. The second section deals with coatings making use of nanotechnology as the enabling technology. These coatings include coatings made from self-assembling smart nanoparticles, nanostructured electro-optically active smart coatings, acrylic nanoparticles, organic–inorganic nanocomposite coatings, and effects of alumina and silica nanoparticles on clear coatings. The last section of the book deals with novel coatings including biocatalytic coatings, superhydrophobic coatings, conducting polymers for intelligent corrosion protection, and protection of steel and aluminum by polyaniline and polyphenylene ether coatings.

We expect that this book will encourage scientific and technological investigators to expand knowledge and technology in this field as well as to apply that knowledge to commercially relevant coatings systems. We thank the authors for their effective oral and written communications and the reviewers for their helpful critiques and constructive comments.

References

1. Joshi, R.; Provder, T.; Kustron, K. "Green Coatings: A Trend is Becoming the Rule Rather Than the Exception". *JCT Coatings Tech.* , **January 2008**, 2–7.
2. Uhlman, P.; Frenzel, R.; Voit, B.; Mock, U.; Szyska, B.; Schmidt, B.; Ondratschek, D.; Gochermann, J.; Roths, K. "Research Agenda Surface Technology: Future Demands for Research in the Field of Coatings Materials". *Prog. Org. Coat.* **2007**, 122–126.
3. *Smart Coatings*; Provder, T.; Baghdachi, J., Eds.; ACS Symposium Series 957; American Chemical Society: Washington, D.C., 2007.

Theodore Provder
Polymer & Coatings Consultants, LLC.
26567 Bayfair Drive
Olmsted Falls, OH 44138

Jamil Baghdachi
Coatings Research Institute
Eastern Michigan University
430 West Forest Avenue
Ypsilanti, MI 48197

Overview

Chapter 1

Smart Coatings

Jamil Baghdachi

Coatings Research Institute, Eastern Michigan University, Ypsilanti, MI 48197

Materials that are capable of adapting their properties dynamically to an external stimulus are called responsive, or smart. The term "smart coating" refers to the concept of coatings being able to sense their environment and make an appropriate response to that stimulus. The standard thinking regarding coatings has been as a passive layer unresponsive to the environment. The current trend in coatings technology is to control the coating composition on a molecular level and the morphology at the nanometer scale. The idea of controlling the assembly of sequential macromolecular layers and the development that materials can form defined structures with unique properties is being explored for both pure scientific research and industrial applications. Several smart coating systems have been developed, examined, and are currently under investigation by numerous laboratories and industries throughout the world. Examples of smart coatings include stimuli responsive, antimicrobial, antifouling, conductive, self-healing, and super hydrophobic systems.

Introduction

Over the past 25 years, coatings technologies have been influenced by the need to lower Volatile Organic Contents (VOC) as well as to reduce the use of costly petroleum based solvents. Waterborne, powder, UV-curable and high solids coatings have had significant growth. Traditionally, these coatings had the primary functions of protecting and decorating substrates.

In addition to VOC reduction, major efforts have been directed to understanding the basic scientific principles that control coating formulation, property enhancement, and its longevity and durability. Among recent performance related studies, investigation of improving acid etch, corrosion, and scratch resistance are noteworthy. Other research and development contributions have been in raw material design, including polymers made through Atom Transfer Radical Polymerization (ATRP) and novel polymer bound additives. Such developments have enabled formulators to design coatings for specific end-uses and well-defined performance requirements.

In the past few years, coatings research has taken a new turn. Nanotechnology is, of course, still the major technology driver in this area. Such influence is mainly due to the development and availability of innovative particle systems, polyelectrolytes, liquid crystals, conductive polymers, as well as nano-structured sol-gel systems. These innovations are enabling the design of novel coatings with exceptional properties and at the same time, allowing the control of the design of the coating more precisely and on a molecular scale.

More recently, there has been growth in research and development and the commercial product generation of coatings which have novel functions. These products sense and interact with their environment in addition to having the traditional protection and decorative functions. These coatings are often referred to as *smart coatings*. More specifically, a smart coating is one which detects changes in its environment, interacts and responds to changes while maintaining compositional integrity. The changes it may respond to include light, pH, biological factors, pressure, temperature, polarity, etc. Hence, a smart coating is tailored in such a way that one or more of the above listed functions can be "switched on" or "switched off" depending on the type and strength of an external signal. Because of such novel switchable functions, these types of coatings generally offer significant added value.

The two potential driving forces for such developments are the need for microelectronic device miniaturization, and multi-functionality as a surface coating. The driving force of nano-revolution is a continuous progress in microelectronics toward increasing the integration level of integrated circuits (IC) and thus the reduction in size of active elements of IC's. In the last decade the size of active elements (e.g., transistors) have been reduced by a factor of two every 18 months (1). Smart coatings are contributing to such developments and can replace mechanical sensors, reduce the number of moving parts, as well as weight and size reduction.

Science and Technology of Smart Coatings

Multifunctional smart coatings have been developed that sense corrosion, pressure, and temperature (2). Additionally, protective and decorative coatings that are self-healing are in commercial use today. Smart coatings are also playing major roles in medical fields by offering permanent antimicrobial and anti-inflammatory medical devices, including implants and release-on-demand medications. Other examples of smart coatings include but are not limited to reversible thermochromic, piezoelectric paint, hydrophilic/hydrophobic switching, self-cleaning, pH responsive, and self-erasing inks.

As mentioned, major efforts have been directed to designing materials that behave predictably and statically. This means that the properties are permanently defined and the behaviors can be reasonably predictable. This, in effect, ensures the longevity of the material during service. Possible changes in the environment of the material during its use are only taking into account (in such a way, that the material disposes of **"reserves"**) to withstand the impact of environmental changes in order to maintain the predefined function. But the property profile is fixed. Here, the "reserves" are viewed as structurally sound and durable polymers such as hydrolytically stable high molecular weight acrylic polymers or certain types of enhancing additives such as ultraviolet absorbers that can inhibit free radical attack and polymer degradation.

High performance conventional coatings use well-designed polymers that take into account the service environment and environmental exposure conditions. For example, an exterior automotive coating must be chip, scratch, corrosion and etch resistant. The chip resistance property is achieved by using tough polymers that dissipate external physical forces and recover quickly upon impact. However, repeated exposure to stone impact gradually reduces the elasticity of the polymer. In this instance, under continuous environmental stimuli, the recoverable elongation shifts toward unrecoverable elongation, yielding a ductile polymer, hence the chip resistance is compromised. This is an example where the reserves are depleted to an extent that the integrity of the system is compromised.

In contrast to the above case, the situation is fundamentally different in nature and biology. Since living beings face continuously new situations, the materials they are made of have to cope with the permanent changes encountered in order to guarantee survival.

In biology, the standard property under variable conditions is assured and maintained by permanent reorganization, rebuilding and reshaping. The growth and healing of bones is an example of such behavior. Alternatively, nature has created materials that change their properties dynamically. These materials communicate with environmental factors and respond to changes by altering their properties and functions to meet the given requirements. Cell membranes are a good example of such responsive materials. Materials which are capable of adapting their properties dynamically to an external stimulus, are called stimuli responsive or "smart materials." Hence, **smart coating is a coating which**

detects and responds dynamically to changes in its environment in a functional and predictable manner. Many so called smart coatings that do not respond to changes in a dynamic and reversible manner may actually be classified as very high performance and novel coatings.

Smart Coatings can be designed and prepared in many ways such as by incorporating stimuli responsive materials such as light, pH, pressure, temperature, etc. sensitive molecules, nano-particles and antimicrobial agents as additives, or by strategically designing polymer structures and coatings that respond to either internal or external stimuli.

In principle, in order to obtain **responsiveness**, two actions must happen concomitantly as well as selectively (3):

- a stimulus must be initiated and the signal be received by the material
- a chemical or physical process is induced by this signal

In order for the response to be predictable and functional, the stimulus must be clear, specific and unambiguous to ensure that the resultant responses do not interfere or cancel each other all together. Those signals that function by triggering responses within the coating itself and with the aim of modifying the bulk properties are called **internal stimuli** such as those in corrosion sensing or self-healing coatings. Responses that alter the surface characteristics relative to the environment such as in self-cleaning coatings are **external stimuli**.

Signals can either be **momentary** or **continuous**. In the case of a momentary signal, a burst of stimulus just long and strong enough is needed to switch the properties of the material from one state to another. Hence, the material will remain in an altered state until an opposing signal reverts the properties back to the original state. For example, materials that respond to pH changes will require (higher or lower pH) to return to the original state. Smart coatings responding to such stimuli are more challenging to create because the state with modified properties must be fairly stable. In the case of continuous stimuli, the modified properties remain unchanged as long as the signal persists (e.g., pressure or temperature change).

Signal(s) acting on the system may produce a smart and unique behavior which may remain permanently fixed, thus not allowing it to return to the original state under any circumstances. Two clear examples of such unidirectional systems include, self-healing and antimicrobial coatings. Due to availability of monitoring tools, measurement and the convenience of fabrication of unidirectional systems, more of these have been carried out and many more are underway. On the other hand, true two-directional system must be able to switch repeatedly from one state to another and perhaps thousands of times during its service life. Examples of such materials, although rare, are thermochromic and pressure sensing coatings.

Since the signal must work on some organic or inorganic chemical entity (polymer, pigment, additive, etc.,) the stimulus can either be physical or chemical in nature. The physical stimuli can include, light, temperature, electrical field, solubility, acoustic and electromagnetic waves, pH, ionic strength, pressure, electrical and surface tension gradients. The physical signals are abundant, more tunable and perhaps less complicated than the chemical counterparts. The chemical stimuli include acid-base, photochemical, electrochemical, redox, and biochemical reactions as well as chemical bond formation and breakings. While there are multiple chemical reactions that can be used as stimuli, monitoring the extent of reactions, their levels and limits are far more difficult and complex.

Structural or property changes are the ultimate result of stimuli responsive materials or smart coatings. For example, structural or configurational change responsible or color change of a reversible simple system containing diazobenzene is limited to *cis-trans* conformation, while minimal structural change is observed in a silver containing antimicrobial coating.

Examples of smart coatings

Modification of surfaces by chemical and physical means to regulate adhesion, adsorption, wettability, etc., is a well-known and widely used approach (4). The ability to reversibly switch the properties of a solid surface from a strongly hydrophobic to a strongly hydrophilic has been demonstrated by grafting various polymers onto polymeric and non-polymeric solid materials (5). Schematic of surface modification and its response to solvent initiated morphology is shown in Figure 1.

In similar investigations brush-like structures containing polystyrene and poly (2-vinylpyridine), P2VP, were exposed to various agents (stimuli) and the response in wetting properties were measured. For example, after exposure to toluene the surface became hydrophobic and the outer top surface contained polystyrene brushes while after exposure to hydrochloric acid produced a hydrophilic surface containing predominantly polyvinyl pyridine (6).

The well-known pH indicator phenolphthalein changes from colorless to red as the pH rises. This phenomenon has been applied to detect the corrosion of aluminum (7). While reversible systems based on pH –responsive acid-base reactions are more abundant and quite easily accessible, in practice their utility is hampered by their lack of repeatability for more than a few times as salts accumulate over time. In addition, such reactions are sensitive to counter ions found in most coating formulations (8). Examples of acid and base functional polymers are shown in Figure 2.

Acid-based reactions that respond to pH variations have also been used to demonstrate switching behavior of brush-like polymers. To demonstrate such responsiveness, polyacrylic acid (PAA) and poly(2vinylpyridine) polyelectrolyte brushes were grafted to a silicone wafer (9). It was demonstrated that at high and low pH values, Figure 3, the top of the sample is occupied by hydrophilic protonated P2VP and dissociated PAA respectively, therefore, the brush remains hydrophilic in the entire range of pH, except in the neutral region, where a compensation of the charges takes place. In pH 2, PAA brush is hydrophobic

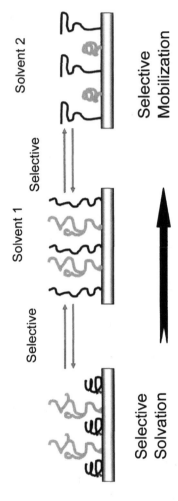

Figure 1. Schematic representation of a solvent initiated morphology change.

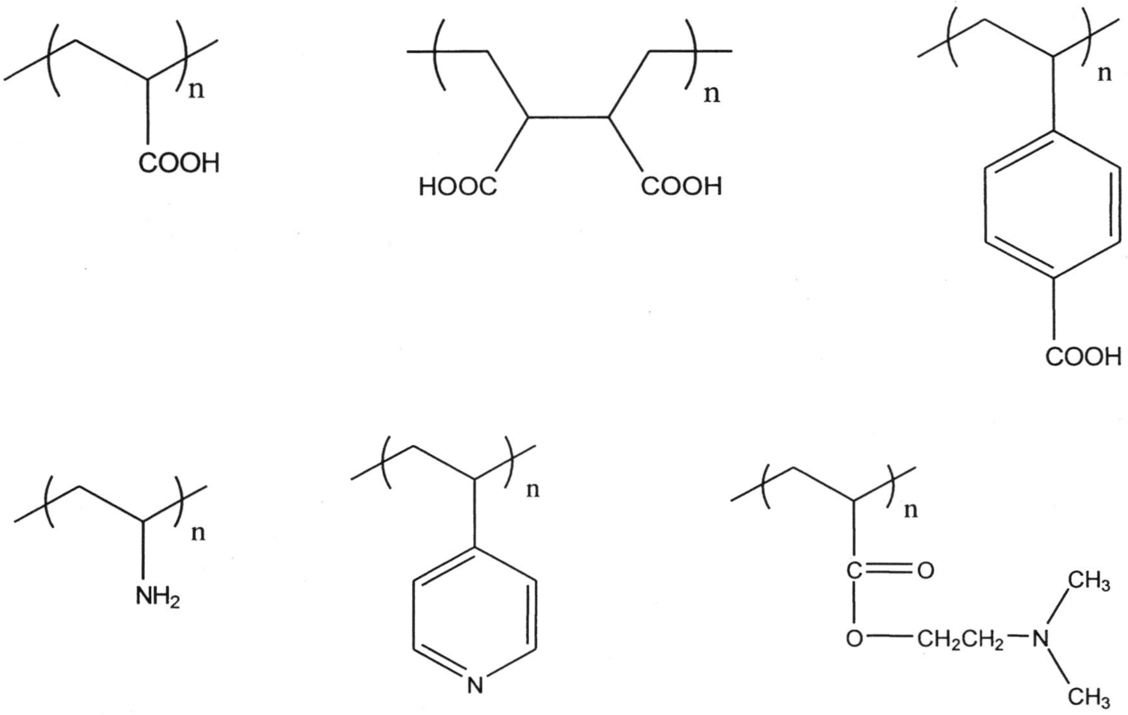

Figure 2. Representative chemical structures of acid and base functional polymers.

with a contact angle of 70°, where at pH 10, it behaves hydrophilic with a contact angle of about 20°, while a P2VP brush demonstrates an inverse behavior. Responsive behaviors of such systems are of potential interest for drug delivery systems and smart nano-devices.

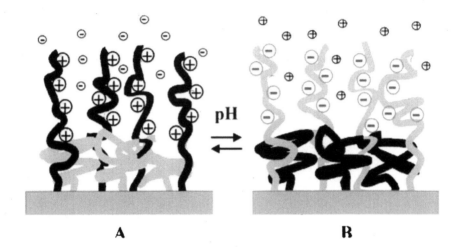

Figure 3. Schematic representation of switching behavior of mixed PE brush upon change of pH: below isoelectric point (A) and above the isoelectric point(B).

Among systems responding to redox reactions, two examples of polyaniline and polythiophene are noteworthy (10). One major advantage of these systems is that the redox reactions produce colored substances.

Photochemical reactions offer a practical approach to reversible systems. As mentioned above, a more stable *trans*-isomer of azobenzene containing coatings (11-12) upon irradiation converts to a *cis*-isomer which can be reverted to the original state by the application of heat and light. Two other systems that are capable of producing photochemical reversible systems are shown in Figure 4.

Organic and inorganic thermochromic pigments and polymers offer a convenient way for designing smart coatings. Certain conjugated polymers and cholesteric liquid crystalline polymers are well-established classes of single component polymers that exhibit thermochromic properties. In some liquid crystalline polymers, under special circumstances, the annealing of the photo-oriented films above the glass transition temperature within the mesophase results in a strong amplification of the optical anisotropy.

Organic dyes can also be incorporated into polymeric systems that switches color as a function of temperature stimuli. For example, a three component

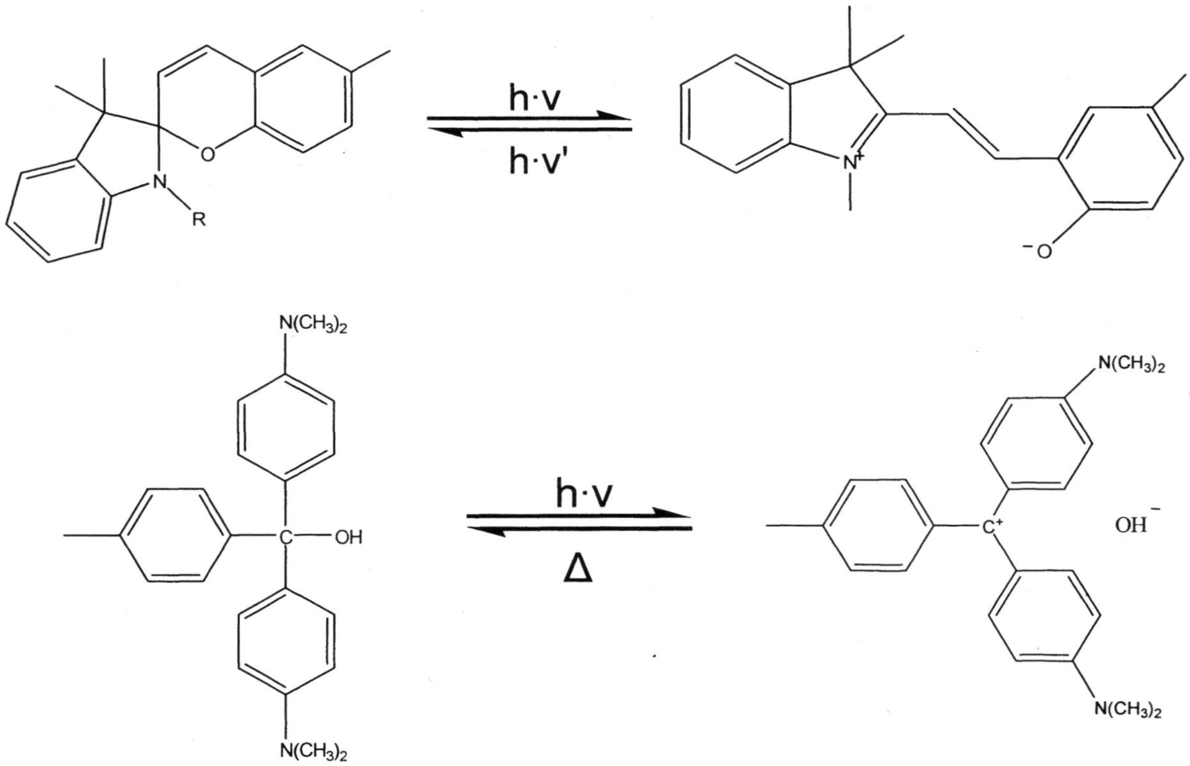

Figure 4. Example of suitable redox switches

system made of a leuco dye, Figure 5, a developer and a sensitizer shows reversible thermochromism in which the colorless dye is transformed to a colored form by heating in a reversible manner.

As shown in Figure 6 (a), the absorption spectra of such a thermochromic system with polyethylene film exhibits an absorption band in the visible range at about 617 nm at 25 °C that disappears when the sample is heated to 80 °C. The thermochromic behavior is illustrated in Fig 6 (b) in which the blue film at ambient temperature switches to colorless upon dipping in hot water (13).

Reversible thermochromic materials are widely used as safety devices. High temperature, reversible indicators give visible indication that a surface is hot. They are used on radiator caps in automobiles and fire-resistant doors to prevent burns. In addition to industrial applications, the technology has been applied to kitchenware in the form of saucepan handles which change to a red color when they reach about 45 °C.

Photochromic coatings which darken reversibly on exposure to light have already found commercial applications. Most photochromic coating materials are based on organic photochromic materials such as spirooxazine, and pyrans, incorporated into organic polymeric or inorganic composites by a sol-gel process. These coatings with response times of about seconds are applied onto transparent substrates (glass, polymers) or on non-transparent substrates (ceramics, polymers, paper) with dry film thicknesses in the range of 5-50 μm.

Smart window coating materials are characterized by their ability to adjust to light transmission upon application of an electrical potential. Materials used in smart window devices include suspended particles, liquid crystals and electrochromics (14).

Electrochromic materials can change their color when a potential is applied due to electrochemical oxidation and reduction. Both inorganic transition metal oxides such as WO_3 and organic polymers such as [3,3-dimethyl-3,4-dihydro-2H-thieno[3,4-b][1,4]dioxepine which requires indium tin oxide and vanadium oxide counter layers are employed (15, 16).

Pressure sensing paints (PSP) are comprised of a luminescent dyes and metal complexes such as divalent osmium complexes (17). Ruthenium (18-20).Iridium (21-24), as well as platinum (25-30) as the luminophore. The phosphorescent dyes are usually dissolved into, or are part of a polymer matrix. The sensing functions by oxygen quenching of the excited state; hence, emission intensity varies with changes in oxygen concentration or pressure. When molecules of the above agents absorb quanta of energy they are promoted into an excited state. From this excited state, there are a number of pathways for the molecule to return to the ground state. All of the pathways have consequences as to how the PSP performs. The processes of luminescence and the bimolecular quenching are pathways that allow a PSP to function.

Research and development in unidirectional stimuli responsive coatings have also produced commercially available coatings. Antimicrobial and anti-inflammatory coatings are being used on medical devices (drug eluting), military clothing and on appliances and hospital equipment. Self-healing

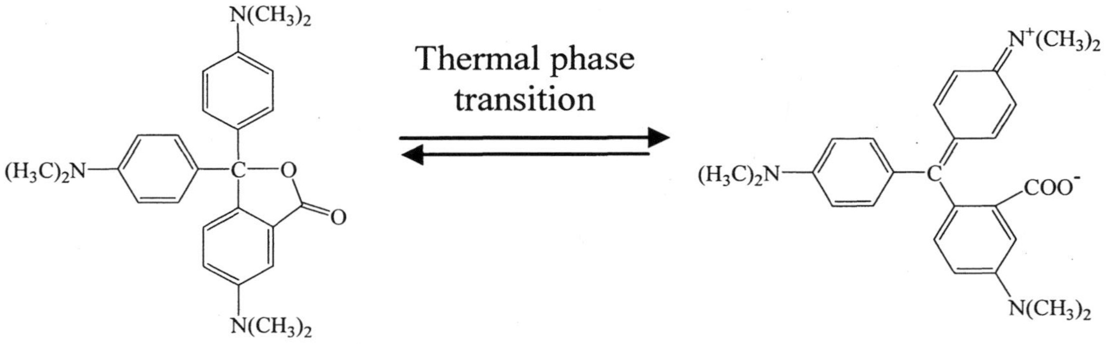

Figure 5. Thermal transition of a colorless leuco dye via ring opening reaction.

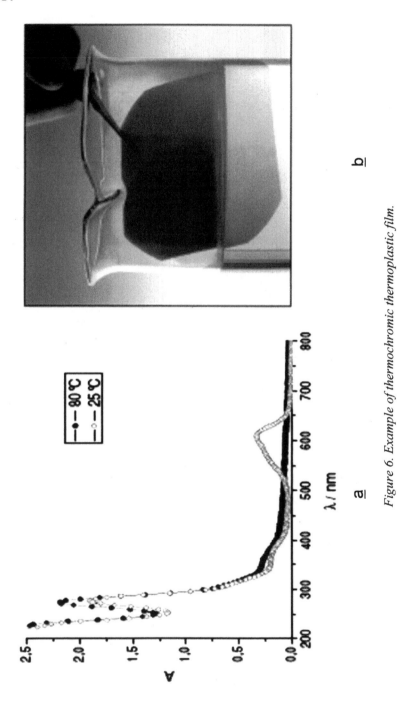

Figure 6. Example of thermochromic thermoplastic film.

coatings, corrosion sensing coatings, radio frequency identification coatings, intumescent coatings, as well as self-erasing inks are examples other unidirectional responsive coatings.

In general, antimicrobial coatings are made either by incorporating biocidal agents into formulation, or polymer structures (31, 32). All of the four classes disrupt the microorganisms' usual activity thereby, rendering it inactive (33). Silver and its various derivatives have been used extensively. (34-42).

Silver containing antimicrobial coatings are made by dispersing a few percent of silver particles either in free form as nano-particles or encapsulated with various porous materials in the polymeric matrix. The antimicrobial activity in these systems arise from the ion exchange, i.e., exchanging ions necessary to cell metabolism with unnecessary silver ion, resulting in disruptions and gradual death (43).

A fair amount of research work and testing along with application has been done with quaternary amine salts. Quaternary ammonium compounds (QAC's) are some of the most commonly used antimicrobials. Common characteristics among QAC's are that they possess both a positive charge and a hydrophobic segment (44-51).

Classification and biological activity of QAC's depend upon the nature of the organic groups attached to nitrogen, the number of nitrogen atoms present, and the counterion (50).

Halamine has been used to coat and produce inherently antimicrobial fibers and surfaces. Halamine must be activated by the addition of chlorine atoms, such as simple household bleach (52). A schematic structure of a recent polymeric system containing halamine is shown in figure 7.

Titanium dioxide (TiO_2), is present in three crystalline polymorphs: rutile, anatase and brookite. Anatase crystal is the most photoactive. Cancer cells have been demonstrated to be susceptible to the chemistry of photocatalysis (53). There have been a plethora of studies published on the use of titanium dioxide as a photocatalyst for the decomposition of organic compounds (53-57). Typically the anatase form is utilized in this application, as it is the most photoreactive.

When the TiO_2 photocatalyst is exposed to light in the presence of water vapor, hydroxyl radicals and superoxide anions are formed. Titanium dioxide coatings containing pigment grade or nano-sized particles, therefore, have been used as antimicrobial coatings.

Concern with any additive approach is the loss of active ingredients due to leaching out and migration. To remedy this, various polymeric systems (Figure 8) containing tethered quat amines have been used.

Smart antimicrobial coatings may also be used to coat medical implants that are suspected to cause infection, tissue destruction, premature device failure or the spread of infection to other areas (58-61).

Coatings that can release antibiotics on demand are quite interesting. Many antibiotic-treated medical devices release antimicrobials continuously at high

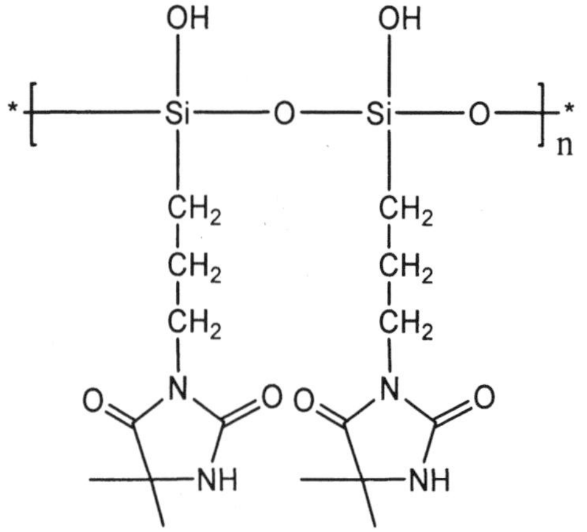

Figure 7. Antimicrobial agent based on halamine derivatives (52).

dose levels, irrespective of whether infection is present or not. An ideal drug delivery system releases a drug at specific locations and at the required time. Investigations are underway to design a new antibiotic delivery system which could accelerate the delivery of drug containing products under conditions present during an infection. The implantation of a synthetic biomaterial elicits a number of biological responses, one of which is inflammation (62). The polymeric drug delivery system being investigated takes advantage of the enzymes released during the inflammatory process in order to trigger the degradation of the polymer. Subsequently, antibiotic drugs are released which were originally used as actual monomers in the synthesis of the polymer itself. Since the inflammatory response is activated in the presence of bacteria or injury, which may predispose a patient to infection (58), an antibiotic could be released, proportionate to the magnitude of the response. Since polyester-urethanes have been shown to be susceptible to degradation by inflammatory enzymes (63), a polymer containing similar chemistry is targeted as an initial model for this investigation.

Inspired by biological systems such as cells and bones in which damage triggers restructuring, reshaping, rebuilding and healing, structural polymeric materials have been developed that possess the ability to self-heal. In one approach microencapsulated healing agents and chemical catalysts are dispersed in an epoxy coating layer (64-66). When damage occurs in the polymer, a crack propagates through the matrix rupturing the microcapsules in the crack path. The ruptured microcapsules release the healing agent which is then drawn into the crack through capillary action. Once the healing agent within the crack plane comes into contact with the embedded catalyst, a chemical reaction is triggered and polymerization of the healing agent occurs.

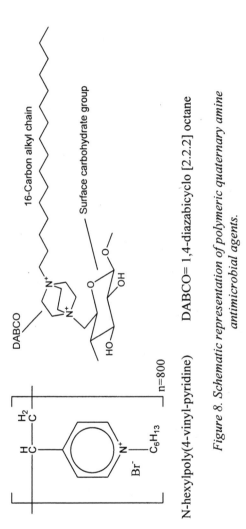

16-Carbon alkyl chain

Surface carbohydrate group

DABCO

n=800

N-hexylpoly(4-vinyl-pyridine)   DABCO= 1,4-diazabicyclo [2.2.2] octane

Figure 8. Schematic representation of polymeric quaternary amine antimicrobial agents.

18

Yet another approach (68), takes advantage of ever existing ordinary stimuli in the environment. Most coatings are exposed to atmospheric elements of water, various electromagnetic radiations and a wide range of temperatures. The natural process of mechanical, hygrothermal fatigue and chemical factors degrade coating matrix and initiate mirocracks within a coating. Propagation of microcracks may result in fracture, loss of barrier properties, delamination and coating failure. Schematic representation of coating degradation upon exposure to atmospheric elements and subsequent self-healing is demonstrated in Figure 9.

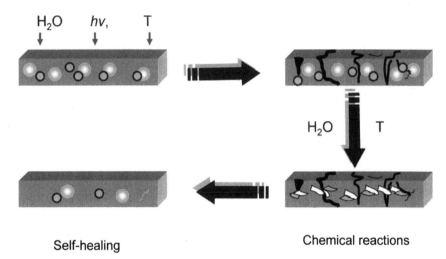

Figure 9. Schematic representation of self-healing sequences in a polymeric coating.

In this approach, microencapsulated healing agents are dispersed in the coating medium. The capsules are designed to rupture and dissolve in water or once exposed to high temperatures. Upon exposure to harsh and abusive conditions such as high temperature and high humidity and mechanical forces, the microcapsules dissolve, melt or rupture and release the healing agents. The reactants are then drawn into the crack through capillary action. Chemical reactions between healing agents take place triggering on-site polymerization and healing of the damaged area. The SEM images of encapsulated healing agents and the cross section of coating containing healing agents are shown in Figure 10.

The healing of the coating is demonstrated by comparing the stress-strain behaviors of exposed and unexposed samples of the coatings. As can be seen from the dynamic mechanical analysis curves in Figure 11, the sample containing self-healing agents responds favorably to exposure conditions by

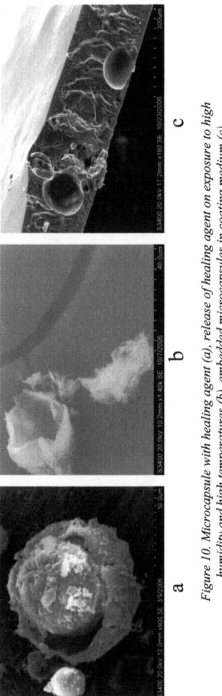

Figure 10. Microcapsule with healing agent (a), release of healing agent on exposure to high humidity and high temperatures (b), embedded microcapsules in coating medium (c)

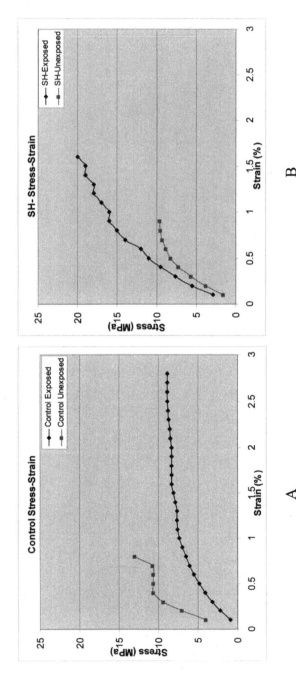

Figure 11. Control without healing agent (A), -■- Control unexposed, -◆- Control exposed at 75% RH, 45 °C; sample with healing agent (B), -■- SH-unexposed, -◆- SH-exposed.

increase in modulus, while the control (without healing agent) is apparently plasticized by exposure to water and high temperatures.

A similar trend is also observed when the free-film samples of a control and a self-healing coating are tested according to ASTM D638, tensile properties of plastics and ASTM D1474, indentation hardness test methods. As can be seen from Table I, the average tensile value of the exposed polyurethane coating increases about 5%, presumably due to the reaction of unreacted isocyanate groups accompanied by an expected increase in elongation and reduction in coating hardness due to plasticization by water. On the other hand, exposure of the self-healing coating results in an increase of about 25%, accompanied by negligible increase in elongation and the fair amount of increase in surface hardness.

Table I. Tensile, elongation and surface harness comparison of control and self-sealing coating.

Sample	Tensile (Psi)	Strain %	KHN, Hardness
Control. Unexposed	31.29	14.75	3.38
Control. Exposed*	33.05	16.91	3.32
SH55 Unexposed	32.12	13.77	3.56
SH55 Exposed	40.34	15.13	4.98

* 24 hours at 42 °C, 75% RH

Based on the above preliminary results, it seems obvious that the self-healing property can restore the integrity of a coating which has developed a minor crack as a result of simulated environmental aging.

Conclusion

Smart coatings have been around for much longer than the term itself. These novel coatings offer technological capabilities which conventional coatings cannot. It seems clear from this review that many of the approaches which are now simply interesting research projects will find their way into mainstream coatings applications in the near future.

22

References

1. Nabok, A. *Organic and inorganic Nanostructures* **2005**, Artech House, MA, USA.
2. Shen, Y.; et al, *Polymer Preprints* **2002**, *43*(1) 69-70.
3. Laschewsky, A.; Lotsch, D.; Seeboth, A.; Storsberg, J.; Stumpe, J. *Smart Coatings III* **2004** European Coatings, Vincentz.
4. Minko, S.; Motornov, M.; Eichhorn, K.; Nitschke, M.; Simon, F.; Stamm, M. *Langmuir* **2003**, *19*, 8077-8085.
5. Minko, S.; Müller, M.; Motornov, M.; Nitschke, M.; Grundke, K.; Stamm, M. *J.Am. Chem. Soc.* **2003**, *125*, 3896-3900.
6. Minko, S.; Motornov, M.; Eichhorn, K.; Nitschke, M.; Simon, M.; Stamm, M. *Langmuir*, **2003**, *19*, 8077-8085.
7. Zhang, J.; G. Fenkel, J. *paper presented at the Material research Society meeting, Boston*, **1997**.
8. Laschewsky, A.; Lotsch, D.; Seeboth, A.; Storsberg, J.; Stumpe, J. *Smart Coatings III*, **2004** European Coatings, Vincentz.
9. Minko, S.; Houbenov, N.; Stamm, M. *Macromolecules*, **2003**, *36*, 5897-5901.
10. Anton, P.; Heinze, J.; Laschewski, A. *Langmuir*, **1993**, *9*, 77-85.
11. Anton, P.; Laschewski, A.; Ward, M. *polym. Bull.* **1995**, *34*, 331.
12. Laschewski, A.; Ward, M. *polymer*, **1991**, *32*, 146.
13. Pieroni, O.; Ciardelli, F. *Trends in polym. Sci.*, **1995**, *3*, 282.
14. Laschewsky, A.; Lotsch, D.; Seeboth, A.; Storsberg, J.; Stumpe, J. *Smart Coatings III*, **2004** European Coatings, Vincentz.
15. Xu C.; Liu L.; Legenski S.; Ning D.; Taya M. *J. Mater. Res.*, **2004**, *19*(7), 2072.
16. Welsh D. M.; Kumar A.; Meijer E. W.; Reynolds J. R.; *Advanced Materials,* **1999**, *11*(16), 1379.
17. Xu C.; Tamagawa H.; Uchida M.; Taya M.; *Proceeding of SPIE.* **2002**, 4695, 442.
18. Carlson, B.; Phelan, D. *Smart Coatings,* **2005** Proceedings, Coatings Research Institute, Ypsilanti, Michigan, USA.
19. Puklin, E.; Carlson, B.; Gouin, S.; Costin, C.; Green, E.; Ponomarev, S.; Tanji, H.; Gouterman, M., *J. App. Poly. Sci.* **2000**, *77*(13), 2795.
20. Ji, H-F.; Shen, Y.; Hubner, J. P.; Carroll, B. F.; Schmehl, R. H.; Simon, J. A.; Schanze, K. S., *Appl. Spect.* **2000**, *54*(6), 856.
21. Gouin, S.; Gouterman, M., *J. App. Poly. Sci.* **2000**, *77*(13), 2815.
22. Vander Donckt, E.; Camerman, B.; Hendrick, F.; Herne, R.; Vandeloise, R., *Bull. Soc. Chim. Belg.* **1994**, 103, 207.
23. Amao, Y.; Ishikawa, Y.; Okura, I., *Anal. Chim. Acta,* **2001**, *445*, 177.
24. Gao, R.; Ho, D. G.; Hernandez, B.; Selke, M.; Murphy, D.; Djurovich, P. I.; Thompson, M. E., *J. Am. Chem. Soc.* **2002**, *124*, 14828.

25. DeRosa, Maria C.; Hodgson, Derek J.; Enright, Gary D.; Dawson, Brian; Evans, Christopher E. B.; Crutchley, Robert J., *J. Amer. Chem. Soc.* **2004**, *126*(24), 7619.

26. Jiang, F.; Xu, R.; Wang, D.; Dong, X.; Li, G., *Gongneng Cailiao* **2000**, *31*(1), 72.

27. Schanze, K. S.; Carroll, B. F.; Korotkevitch, S., Morris, M., *AIAA J.* **1997**, *35*(2), 306.

28. Jiang, F-Z.; Xu, R.; Wang, D-Y.; Dong, X-D.; Li, G-C.; Zhu, D-B., Jiang, L., *J. Mat. Res.* **2002**, *17*(6), 1312.

29. Sakaue, H.; Gregory, J. W., Sullivian, J. P., *AIAA J.* **2002**, *40*(6), 1094.

30. Bowman, R. D.; Kneas, K. A.; Demas, J. N.; Periasamy, A., *J. of Microscopy,* **2003**, *211*(2), 112.

31. Schanze K. S.; Carroll B. F.; Korotkevitch S.; Morris M., *AIAA J.* **1997**, 35(2), 306.

32. Tiller, J. C.; Lee, S. B.; Lewis, K.; Klibonov, A. M. *Biotechnol Bioeng* **2002**, *79*, 465-471.

33. Lin, J.; Tiller, J. C.; Lee, S. B.; Lewis, K.; Klibanov, A. M. *Biotechnol Lett.* **2002**, *24*, 801-805.

34. Donelli G.; Francolini, I; *J. Chemotherapy,* **2001**, *3*(6): 595-606.

35. Klasen, H..J., *Burns,* **2000**, *26*, 131-138.

36. Bromberg, L.E., V.M. Braman, D.M. Rothstein, P. Spacciapoli, S.M. O'Connor, E.J. Nelson, D.K. Buxton, M.S. Tonetti, and P.M. Friden. *J. Controlled Rel.* 2000, 68, 63-72.

37. Quintavalla, S., and L. Vicini, *Meat Science* **2002**, 62:373-380.

38. Wright, J.B., K. Lam, D. Hansen, and R.E. Burrell. *Am. J. Infect. Control* **1990**, *27*:344-350.

39. Clement, J.L. and P.S. Jarrett. *Metal-Based Drugs* **1994**, *1*, 467-482.

40. Thurman, R.B. and C.P. Gerba. *CRC Critical Reviews in Environmental Control* **1989**, 8, 295-315.

41. Yahya, M.T., L.K. Landeen, M.C. Messina, S.M. Kute, R. Schulz, and C.P. Gerba. *Can. J. Microbiol.* **1990**, *36*, 109-116.

42. Becker, R.O. *Metal-Based Drugs* **1999**, *6*, 311-314.

43. Maki, D.G., and P.A. Tambyah, Emerg. *Infect. Dis.,* **2001**, 7:1-6.

44. Feng, Q. L., Wu, J., Chen, G.Q., Cui, F.Z., Kim, T. N., Kim, J.O, *J. Biomed Mater Res,* **2000**, *52*, 662-668.

45. Tiller, J.; Lee, S.; Lewis, K; Kilbanov, A., *Biotechnology and Bioengineering* **2002**, *79*, (4) 465-471.

46. Li, G. J.; Shen, J. R.; Zhu, Y. L. *J Appl Polym Sci.* **2000**, *78*, 668.

47. Gabrielska, J.; Sarapuk, J.; Przestalski, S.; Wroclaw, P. Tenside, *Surfactants, Detergents,* **1994**, *31* (5), 296.

48. Robertson, J. R. Eur. Pat. 0,611,782, A1, **1994**.

49. Talaro, K.; Talaro, A., *Foundations in Microbiology*; WCB Publishers: Dubuque, IA, 1993; pp. 286.

24

50. Tiller, J. C.; Liao, C. J.; Lewis, K.; Klibanov, A. M. *Proc Natl Acad Sci.* **2001**, *98* (11), 5981.
51. Li, G. J.; Shen, J. R.; Zhu, Y. L. *J Appl Polym Sci.* **2000**, *78*, 668.
52. Cohen, J. I.; Abel, T.; Filshtinskaya, M.; Melkonian, K.; Melkonian, A.; Burkett, D.; Engel, R. Abstracts of Papers, 223rd ACS. National Meeting; Orlando, FL, 2002; CARB-059.
53. Lin, J.; Winkelmann, C.; Worley, D.; et al., *J. Appl. Polym. Sci.* **2002**, *85*, 177-182.
54. Sunada K., Watanabe T., and Hashimoto K., *Environ. Sci. Technol.* **2003**, *37*, 4785-4789.
55. Blake, D.M., P. Maness, Z. Huang, E.J. Wolfrum, and J. Huang. *Bull. Environ. Contam. Toxico.* **1999**.
56. Matthews, R. W., *J. Chem. Soc. Faraday Trans.* I. **1989**, *85*:1291.
57. Carey, J. H., J. Lawrence, and H. M. Tosine. *Bull. Environ. Contam. Toxicol.* **1976**, *16*:697.
58. Turchi, C. S., and D. F. Ollis. J. Catal. **1989**, *119*, 483.
59. Gristina AG. *Science*, **1987**, *237*, 1588.
60. Dankert J, Hogt A.H, Feijen J., Biomedical polymers, CRC Crit Rev Biocomp, **1986**, *2*, 219-301.
61. Buret A, Ward KH, Olson ME, Costerton JW., *J. Biomed Mater Res* **1991**, *25*, 865-874.
62. Dickinson GM, Bisno A.L., *Antimicrob. Agents Chemother.* **1989**, *33*:597.
63. Anderson J.M., *Cardiovasc. Pathol.*, **1993**, *2*, 334.
64. Santerre JP, Labow RS, Duguay DG, Erfle D, Adams GA. *J. Biomed. Mater Res* **1994**, 28, 1187-1199.
65. White, S.R., Sottos, N.R., Geubelle, P.H., Moore, J.S., Kessler, M.R., Sriram, S.R., Brown, E.N., and Viswanathan, S., *Nature.* **2001**, *409*, 794.
66. Brown, E.N., Sottos, N.R., White, S.R. *Experimental Mechanics* , **2002**, *42*, 372.
67. Kessler, M.R., Sottos, N.R., White, S.R. *Composites Part A.*, **2003**, *34*, 743.
68. Unpublished report.

Bioactive Coatings

Chapter 2

Novel Antibacterial Polymers

Bekir Dizman[1], Mohamed O. Elasri[2], and Lon J. Mathias[1,*]

Departments of [1]Polymer Science and [2]Biological Sciences, The University of Southern Mississippi, 118 College Drive 10076, Hattiesburg, MS 39406–0076

The research presented in this review covers the areas of low molecular weight antibacterial agents and antibacterial polymers. The review is divided into two main sections. In the first section, a detailed background information is provided about low molecular weight antibacterial agents (quaternary ammonium compounds and norfloxacin), antibacterial polymers with quaternary ammonium compounds and norfloxacin, and antibacterial activity tests pertinent to the research. In the second section, the syntheses and antibacterial activities of several new antibacterial polymers are discussed. These polymers are either acrylate/methacrylate or acrylamide/methacrylamide-type polymers with pendant biocidal groups, i.e. mono- and bis-quaternary ammonium compounds or norfloxacin. In general, two approaches were utilized to obtain the antibacterial polymers. The first approach involved the incorporation of the antibacterial agents to monomers, followed by their polymerization. The second approach, on the other hand, involved the linking of the antibacterial agents directly onto preformed functional polymers. The water-soluble and water-insoluble polymers were synthesized according to the pendant groups and comonomers present. All polymers were tested for antibacterial activity against *Staphylococcus aureus* and *Escherichia coli*, representatives of Gram-positive and Gram-negative bacteria, by using broth dilution and shaking flask methods.

Introduction

In the last two decades, there has been a growing global concern about the risks of bacterial contamination. To overcome the problems caused by the bacterial contamination, several types of antibacterial agents have been developed. Although low molecular weight antibacterial agents have been used more commonly, there is an increasing trend in using the antibacterial polymers in recent years. In this review, low molecular weight quaternary ammonium compounds and norfloxacin as well as the polymers containing these antibacterial agents are discussed.

Low Molecular Weight Antibacterial Agents and Antibacterial Polymers

Quaternary ammonium compounds (QACs)

Quaternary ammonium compounds are some of the most commonly used antibacterial agents. Common characteristics among QACs are that they possess both a positive charge and a hydrophobic segment (1,2,3). Classification and biological activity of QACs depend upon the nature of the organic groups attached to nitrogen, the number of nitrogen atoms present, and the counterion (1). QACs usually contain four organic groups linked to nitrogen, which may be similar or different in chemistry and structure. The organic substituents are either alkyl, aryl, or heterocyclic (4). At least one of the organic substituents should be a long alkyl chain in order to provide a hydrophobic segment compatible with the bilayer of the outer cell wall (5,6,7). It has been shown that an increase of the alkyl chain length of an amphiphilic compound, i.e. to 14 carbon alkyl chains, is followed by an increase in the hydrophobic interaction with the lipid bilayer of the cell wall, which in turn increases the antibacterial activity of the compound (7). QACs containing one long alkyl chain substituent of at least eight carbon atoms have been shown to be very active biocides in water (8). The number of nitrogen atoms can vary in the molecule depending on the starting materials used in the synthesis. Both mono and bis-QACs are currently in use. Any anion may be attached to the cation to form a salt, although the chloride and bromide salts are most commonly used (9). Some of the examples of the most commonly used low molecular weight QACs are shown in Figure 1.

QACs are usually white, crystalline powders that are very soluble or dispersible in water. As the chain length of the substituents increases, the solubility of QACs in polar solvents decreases, whereas their solubility in nonpolar solvents increases (10). QACs have a broad spectrum of antibacterial activity and often display extended biological activity because they may leave long-lived residues on treated surfaces (2). They are effective against both Gram-positive and Gram-negative bacteria at medium concentrations, and also have moderate effectiveness against viruses, fungi, and algae (7,11).

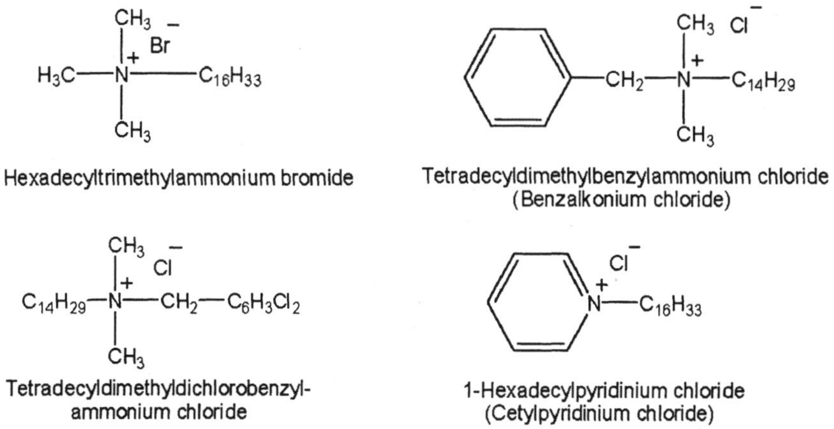

CH₃, Br, H₃C—N⁺—C₁₆H₃₃, CH₃

Hexadecyltrimethylammonium bromide

CH₃ Cl⁻, CH₂—N⁺—C₁₄H₂₉, CH₃

Tetradecyldimethylbenzylammonium chloride
(Benzalkonium chloride)

CH₃ Cl⁻, C₁₄H₂₉—N⁺—CH₂—C₆H₃Cl₂, CH₃

Tetradecyldimethyldichlorobenzyl-
ammonium chloride

Cl⁻, N⁺—C₁₆H₃₃

1-Hexadecylpyridinium chloride
(Cetylpyridinium chloride)

Figure 1. The most commonly used QACs

Some of the advantages of QACs over other antibacterial agents are that they are more stable, less corrosive, non-irritating to the skin, and have low mammalian toxicity (*2*).

As a result, they can be used as antiseptics, bactericides, fungicides, sanitizers, and softeners, but are also used in deodorants and as conditioning agents in hair cosmetics (*12*).

They are also active against human and plant pathogens in the packing house environment, particularly as treatments to sanitize building and equipment surfaces and used as preservatives in ophthalmic and nasal solutions (*13,14*).

Antibacterial Polymers with QACs

Continuous effort has been made during the last two decades to synthesize polymers with QA substituents (*1,3,4,5,15,16,17*). The literature suggests that the polymers containing QACs either in the backbone or as pendant groups show enhanced efficacy over corresponding small molecule QACs plus reduced residual toxicity, increased efficiency and selectivity, and prolonged lifetime (*1,5,18,19*). Polymeric antibacterial agents containing QACs also have the advantage that they are nonvolatile, chemically stable, and do not permeate through the skin. As a result, they significantly reduce losses associated with volatilization, decomposition, and migration (*20*).

Antibacterial activity of polymers with QACs results from physical interactions between polymers and bacteria. These interactions can be based on a bactericidal or bacteriostatic action of polymers against bacterial cells or capture of bacterial cells by the polymers (*17*). Polymers with QACs are favored over small molecule QACs in terms of adsorption onto the bacterial cell wall and cytoplasmic membrane and by insertion into the membrane. Low molecular weight cationic disinfectants function by electrostatic interaction of the antibacterial agent positive charge with the negatively charged species present both in the bacteria cell walls and in the cytoplasmic membranes (*16,17*). The adsorption of polymer with QACs onto the cell surfaces is favored over small molecule cations because of the much higher charge density carried by the polymers (*17*). Cytoplasmic membrane binding is also expected to be facilitated by the polymers, compared to that by small molecule cations, due to the high negative charge density in the cytoplasmic membrane. Consequently, membrane disruption should be enhanced (*5,18*). The effect of electron charge density of different polymers with QACs and their monomers have been studied (*5,18,19*). Ikeda *et al.* have synthesized water-soluble antibacterial polymers with QACs and found that the polymers were more active against Gram-positive and Gram-negative bacteria than the monomers containing QACs in solution (*1,18*). Similar effects were also obtained when polymers with other QACs and biguanide groups were investigated (*5,21*).

Antibacterial activity of polymers with QACs depends on a variety of factors. In addition to the electron charge density, molecular weight of the polymer, polymer concentration in solution, length of the alkyl chains, contact time of the polycations with the bacteria, and type of bacteria also influence the antibacterial activity. The molecular weight dependence of the antibacterial activity of different polymers with QACs and corresponding small cationic molecules has been studied extensively (*5,18*). Increasing the molecular weight from monomers to polymers generally causes an increase in activity. For polymers, however, an increase in the molecular weight leads to an increase in activity only up to a certain extent; an optimal molecular weight range is usually observed for the bactericidal action of the polymers (*18*). Increasing molecular weight increases the charge density of a single coil polyelectrolyte, which leads to enhanced adsorption of polycations onto bacteria cell surfaces. Consequently, more disruption of the cytoplasmic membrane of the bacteria, the target site of the cationic disinfectant, is achieved (*18*). However, there is an upper limit for molecular weight of the polymer because an increase in the molecular weight makes the diffusion of the polymer through the bacteria cell wall more difficult and results in a decrease in antibacterial activity.

The concentration of the polycation also affects antibacterial activity. Disruption of the cytoplasmic membrane is a consequence of interaction between the bound polymers and the membrane; thus, antibacterial activity is expected to be increased with increasing amounts of cationic polymer in solution (*22*). Guangji *et al.* showed that the antibacterial activity of the water-soluble pyridinium-type polymers is enhanced as the total content of pyridinium groups increases by increasing polymer concentration (*17*).

Another factor affecting the antibacterial activity is the length of the alkyl chain attached to the polycation heteroatom. The alkyl chain length of the surfactants influences the extent of the membrane disruption, with higher chain lengths being incorporated more readily into the lipid bilayers of the plasma membrane (*23,24,25*). Cationic disinfectants with chain lengths of 14 to 18

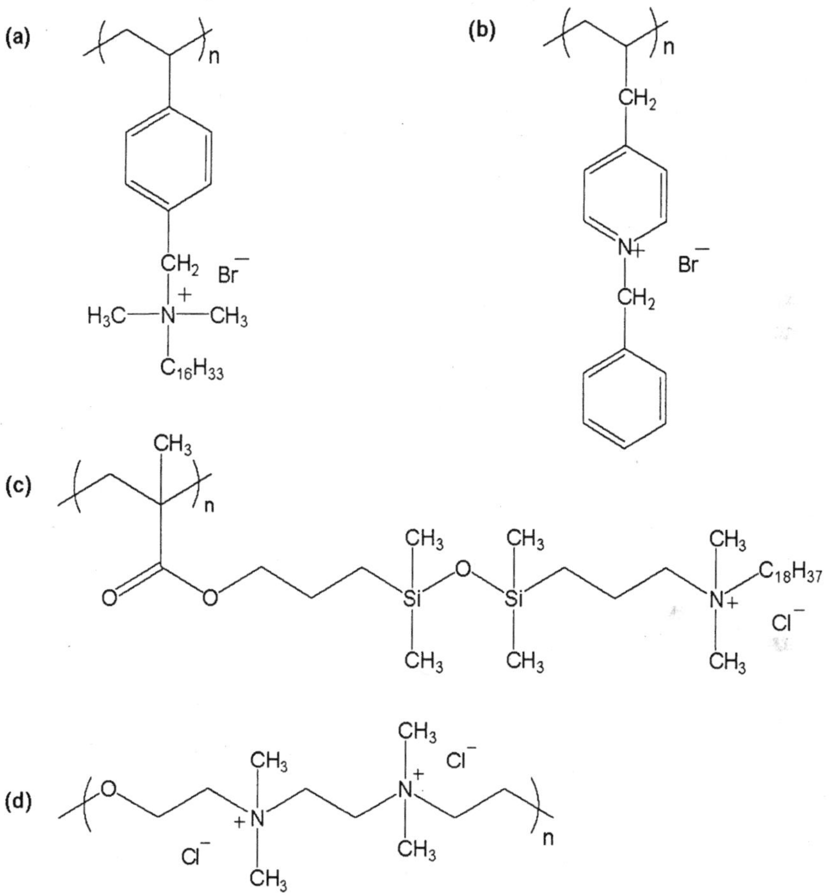

Figure 2. Examples of polymers with QACs

carbons showed maximum antibacterial activity in water against Gram-positive and Gram-negative bacteria (26). However, Klibanov et al. showed that, when attached to a glass surface, polycations with short alkyl chains are more effective against both type of bacteria than the polymers having long alkyl chains (27).

Contact time between antibacterial polymers with QACs and bacteria is also an important factor in antibacterial activity. An increase in the contact time causes an accumulation of dead bacteria around the antibacterial polymers with QACs and decreases their efficiency. After simply washing away dead bacteria, however, polymers continue to be effective (27). It has been observed that the concentration of the dead bacteria must be very high in order to affect the activity (28). A study on ageing has been done to determine the protection lifetime against bacteria. Increases in lipophilicity and the bulkiness of QAC substituents improved the length of the antibacterial activity (29).

Finally, the type of bacteria to which the polymer will be exposed has great influence upon antibacterial activity. In many studies, polymers with QACs have been more effective against Gram-positive bacteria than Gram-negative bacteria (17,18,27). Different activities are due to differences in the cell wall structure between Gram-positive and Gram-negative bacteria (10,27,30,31). The Gram-positive bacteria have a simple cell wall structure and there is only a rigid peptidoglycan layer outside the cytoplasmic membrane. The peptidoglycan layer, although relatively thick, has networks of pores that allow foreign molecules to come into the cell without difficulty (10,32). Gram-negative bacteria, however, have very complicated cell walls. They have an additional outer membrane outside the peptidoglycan layer with structure similar to that of the cytoplasmic membrane. Because of the bilayer structure, the outer membrane is a potential barrier against foreign molecules of high molecular weight (10,33).

The lethal action of low molecular weight cationic biocides is mechanistically complex. Biocide target sites are the cytoplasmic membranes of bacterial cells and the following elementary processes have been identified as modes of action: (1) adsorption onto the bacterial cell surface; (2) diffusion through the cell wall; (3) binding to the cytoplasmic membrane, (4) disruption of the cytoplasmic membrane; (5) release of the cytoplasmic constituents such as K^+ ions, DNA, RNA; and (6) death of the cell (5). The generally accepted mode of antibacterial activity of polymers with QACs is also interpreted by using these six elementary processes (5). Adsorption to negatively charged bacterial cell surfaces is expected to increase with increase in charge density of the cationic biocides. Therefore, it is reasonable to assume that process 1 is enhanced for polymeric cations compared to low molecular weight monomers and model compounds. A similar situation can also be expected in process 3 because there are many negatively charged species present in the cytoplasmic membrane, such as acidic phospholipids and membrane proteins (33). Process 4 is also favored for polymers because the concentration of cationic groups bound to the cytoplasmic membrane will be higher with polymers (33). On the other hand, process 2 will be suppressed as the molecular size of the diffusing species increases, since the thick, rigid peptidoglycan layer of the Gram-positive bacteria acts as a barrier against foreign molecules with high molecular weight (*34*). Consequently, the overall activity will be determined by both polymer favorable (processes 1, 3, and 4) and polymer unfavorable events (process 2) (*20*).

Antibacterial polymers with QACs have been used as coatings in many areas such as food processing (*35*), biomedical devices (*36,37*), and filters (*38*). The use of cationic antibacterial polymers can eliminate bacterial infection of implanted devices such as catheters (*36,37*). These polymers can be used in paints on hospital room walls and everyday objects such as doorknobs, children's toys, computer keyboards, and telephones (*27*). They are used in the textile industry to form antibacterial fibers (*39,40*), and as disinfectants and preservatives in pharmaceuticals (*41*).

Antibacterial polymers containing QACs have been used as cleaning solutions for contact lenses (*42*). One of the problems associated with cleaning soft contact lenses is that they have a high capacity to adsorb water and the compounds employed to disinfect the contact lenses. The disinfectants later may be released when the soft contact lenses are worn on the eye. This may damage or stain the contact lenses and harm the sensitive tissues of the eye. To overcome these problems, antibacterial polymers with QACs may be used to disinfect the lenses. The advantage of using antibacterial polymers is that they have a larger molecular size and are less likely to penetrate or be absorbed into the soft contact lenses. They also tend to be less toxic. These polymers may also be used to construct the contact lenses and impart them with antibacterial activity (*43*). Structures of two of the polymers used for contact lens synthesis and in cleaning solutions for contact lenses are given in Figure 2c and 2d, respectively.

Fluoroquinolones (Norfloxacin)

Another major class of antibacterial agents is fluoroquinolones. They are a group of synthetic antibiotics that exhibit excellent potencies and a broad spectrum of activity against a variety of Gram-positive and Gram-negative bacteria (*44,45,46*). They are widely used in human and veterinary medicine for the treatment of infectious diseases (*47,48*). The antibacterial activity of fluoroquinolones depends on the bicyclic heteroaromatic pharmacophore as well as the nature of the peripheral substituents and their spatial relationship (45). There has been extensive research on the structure-activity relationships of antibacterial fluoroquinolones during the last two decades (*45, 49, 50, 51, 52, 53, 54, 55, 56*).

Norfloxacin is one of the second-generation fluoroquinolone agents (shown in Figure 3a). It shows very good antibacterial activity against various types of bacteria through several mechanisms (*45*). It is a specific inhibitor of DNA gyrase, a bacterial type II topoisomerase, which unwinds the supercoiled DNA prior to replication and transcription (*48*). Norfloxacin also controls DNA superhelicity and plays important roles in various cellular processes such as the efficiency of replication (*57*).

(a)

(b)

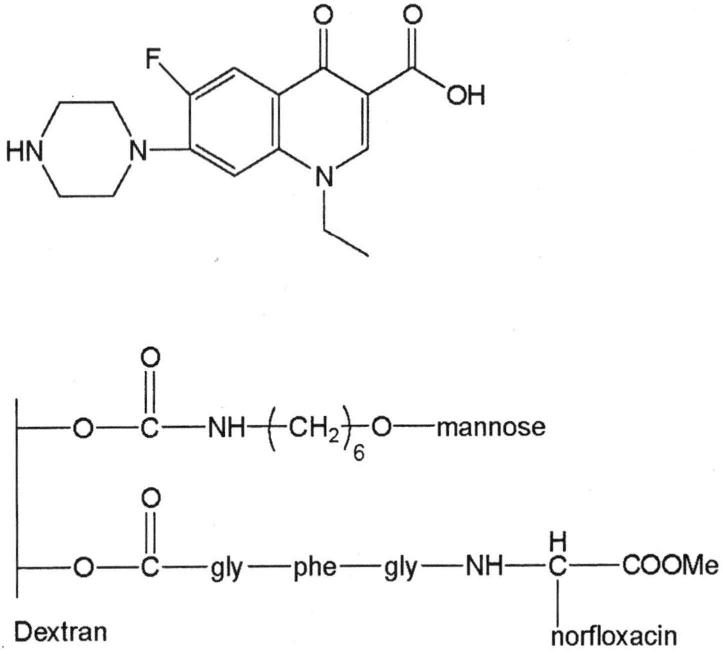

Figure 3. Norfloxacin (a) and polymeric norfloxacin conjugate (dextran-glyphe-gly-gly(α-norfloxacin)-OMe) (b)

Antibacterial Polymers with Norfloxacin

There has been an increasing effort in the last two decades to synthesize polymeric systems that will prevent bacterial infections on implants and biomedical devices. One approach to controlling infection is to incorporate antibiotics within the polymers. Traditionally, incorporation of the antibiotics into polymers is achieved by either grafting or coating drug agents on the surface of polymers by chemical or physical means (*58,59,60,61*), or by physically entrapping the active agent within a polymer matrix or micelle (*62,63*). An alternative approach is to have polymeric carriers incorporating drug moieties either in the backbone of the polymer, or as terminal and pendant groups on the polymer chain.

Recently, norfloxacin was conjugated to mannosylated dextran in order to increase the drug's uptake by cells, enabling faster access and effect on microorganisms (polymer structure shown in Figure 3b) (*64,65*). Results showed that norfloxacin could be enzymatically cleaved from the polymer-drug conjugate (*65*). In another study, the synthesis of polyurethanes having antibacterial norfloxacin drug in the backbone was carried out; their antibacterial activities are currently being tested (*66*). Yoon *et al.* synthesized poly(acrylated quinolone) using norfloxacin as a pendant group on the polymer chain and observed very good antibacterial activities against several Gram-positive and Gram-negative bacteria. They also compounded these polymers with other ordinary synthetic polymers such as low-density polyethylene and poly(methyl methacrylate) and were able to reduce the viable cell number significantly on contact (*67*).

Temperature Responsive Polymers

Stimuli-responsive polymers are polymers that show dramatic property changes in response to small external changes in the environmental conditions. These polymers recognize a stimulus, judge its magnitude, and then change their chain conformation in direct response (*68*). There are many chemical and physical stimuli such as pH, ionic factors, chemical agents, temperature, electric or magnetic fields, and mechanical stress (*68*). Among these stimuli, temperature is the most widely used in stimuli-responsive polymer systems. The change of temperature is relatively easy to control and easily applicable both *in vitro* and *in vivo*. Temperature responsive polymers have a critical solution temperature at which the phase of polymer and solution is discontinuously changed according to their composition. If the polymer solution is homogeneous below a specific temperature and phase-separated above this temperature, the polymer has a lower critical solution temperature (LCST). The term, upper critical solution temperature (UCST), applies to the opposite behavior. Most applications are related to LCST-based polymer systems. The homopolymer of NIPAAm, poly(N-isopropylacrylamide) (PNIPAAm), is one of the most extensively studied polymer and has a LCST of 32 °C at which it undergoes a reversible coil-to-globule transition (*69,70,71,72*). Below the LCST, PNIPAAm is soluble in water; however, above the LCST, the entropic contribution to the free energy dominates the enthalpic contribution from hydrogen bonds, causing the polymer to precipitate. Several copolymers of NIPAAm have been synthesized in order to change the LCST (*73,74,75,76,77,78*). It has been found that the use of hydrophilic comonomers with NIPAAm increases the LCST of the copolymers, whereas the use of hydrophobic monomers decreases the LCST (*70,77*).

Antibacterial Activity Tests

A wide variety of bacterial methodologies are being used by microbiological laboratories around the world. Two of the primary methods are broth dilution (MIC and MBC tests) and shaking flask methods (*79,80,81*). The broth dilution method is a technique in which a standardized suspension of bacteria is tested against varying concentrations of an antibacterial agent (usually with doubling dilutions) in a standardized liquid medium. The broth dilution method can be performed either in tubes containing a minimum volume of 2 mL (macrodilution) or in smaller volumes using microtitration plates (microdilution). The broth dilution method can be used for water-soluble antibacterial agents as well as antibacterial agents that are well dispersed in water. Minimum inhibitory concentration (MIC) values for antibacterial agents can be obtained using this method. MIC is defined as the lowest concentration of the antibacterial agent that prevents the growth of bacteria. If the broth dilution method is followed with the incubation of the solutions in which any growth of bacteria is not observed during broth dilution test on agar plates, minimum bactericidal concentration (MBC) values can be obtained. MBC is defined as the lowest concentration at which all bacteria are killed. At this concentration the growth of bacteria on agar plates is not observed (*79,80*). The MIC and MBC results are very accurate and reproducible.

Another method that is used for testing insoluble antibacterial agents is the shaking flask method (*81*). This is not a standardized technique, although it provides information about the antibacterial activity of the insoluble antibacterial agents. A specified concentration of an insoluble antibacterial agent is mixed with a specified amount of bacteria dispersion and the growth of bacteria in the medium is observed.

Syntheses and Antibacterial Activities of New Polymers

This section discusses the syntheses and antibacterial activities of several new antibacterial polymers. The polymers investigated here are either acrylate/methacrylate or acrylamide/ methacrylamide-type polymers with pendant biocidal groups. The pendant groups are either mono- and bis-QACs or norfloxacin. Two different approaches were utilized to obtain these antibacterial polymers. The first approach involved the introduction of the biocidal groups to monomers, followed by their polymerization. The second approach, on the other hand, involved the post-modification of preformed functional polymers to obtain antibacterial polymers. The synthesized polymers are grouped according to their solubility characteristics in water and the pendant groups that they have. The syntheses and antibacterial activities of water-soluble and water-insoluble polymers with different pendant biocidal groups will be discussed below.

Water-Soluble Methacrylate Polymers with Pendant Bis-QACs

New methacrylate monomers containing pendant bis-QACs based on 1,4-diazabicyclo-[2.2.2]-octane (DABCO) were synthesized. The DABCO group contained either a butyl or a hexyl pendant group comprising the hydrophobic segment of the monomers and one tether group to the methacrylate moiety. The monomers (C4-DAM and C6-DAM) were homopolymerized in water using 2,2'-azobis(2-methylpropionamide) dihydrochloride (V-50) as an initiator (shown in Figure 4). The monomers and polymers were characterized by elemental analysis, NMR, FT-IR, TGA, and DSC (22).

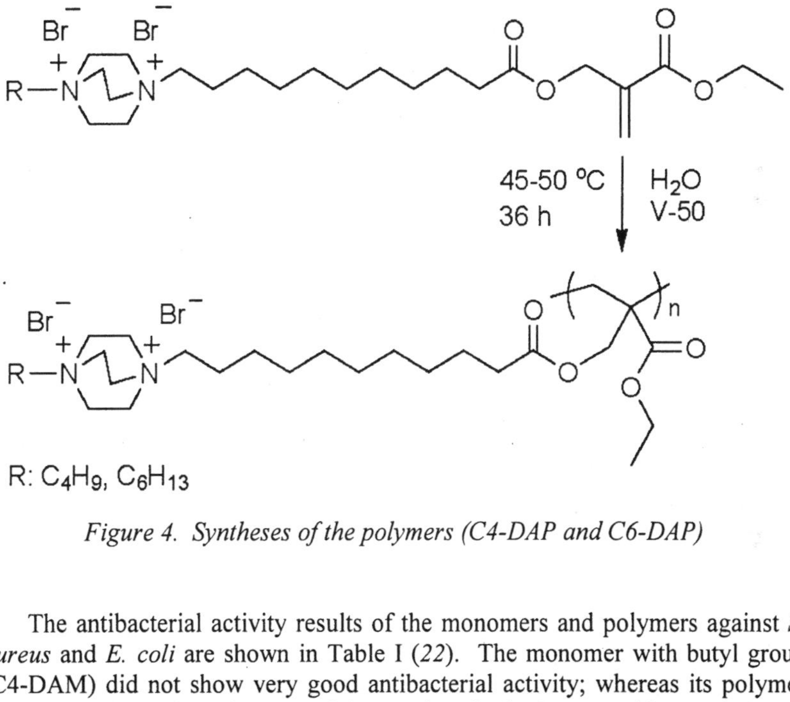

R: C4H9, C6H13

Figure 4. Syntheses of the polymers (C4-DAP and C6-DAP)

The antibacterial activity results of the monomers and polymers against *S. aureus* and *E. coli* are shown in Table I (22). The monomer with butyl group (C4-DAM) did not show very good antibacterial activity; whereas its polymer (C4-DAP) showed moderate activity against both Gram-positive and Gram-negative bacteria. The monomer and polymer with hexyl group (C6-DAM and C6-DAP), on the other hand, had very good and encouraging minimum inhibitory concentration (MIC) values against both types of bacteria. The minimum bactericidal concentration (MBC) values showed a similar trend to the MIC results. It can be concluded from the antibacterial activity test results that the antibacterial activity of both the monomers and polymers increased significantly with an increase of the alkyl chain length from 4 to 6.

Table I. Antibacterial activity results of the monomers and polymers

Monomer/Polymer	Bacteria	MIC (µg/ml)	MBC(µg/ml)
C4-DAM	S. aureus	>256	
C4-DAP	S. aureus	250	250
C4-DAM	E. coli	>256	
C4-DAP	E. coli	250	>1000
C6-DAM	S. aureus	64	
C6-DAP	S. aureus	62.5	62.5
C6-DAM	E. coli	64	
C6-DAP	E. coli	62.5	62.5

Water-Soluble Methacrylate Polymers with Pendant Mono-QACs

New water-soluble methacrylate polymers with pendant mono-QACs were synthesized by reacting the alkyl halide groups of a previously synthesized functional methacrylate polymer (*82*) with various tertiary alkyl amines containing 12, 14, and 16 carbon alkyl chains (Figure 5). The characterization of the functional polymer and the polymers with QACs were carried out by utilizing FTIR, NMR, SEC, TGA, and DSC (*82,83*). The degree of conversion of alkyl halides to QA sites in each polymer was determined by ^{1}H NMR and found to be above 90%.

The antibacterial activities of the polymers with pendant mono-QACs against *S. aureus* and *E. coli* were determined using broth dilution and spread plate methods. The antibacterial activity tests were performed by preparing a range of concentrations of each polymer in water in 96-well microtiter plates and adding the liquid growth medium (tryptic soy broth, TSB) into each well in a water:TSB ratio of 9:1. The addition of TSB resulted in precipitation of the polymers. As a result of the precipitation, the antibacterial activities of the polymers might be limited since there is less interaction between the polymers and bacteria when the polymers are solid compared to the case when they are in solution. Although the precipitation of the polymers was present, the polymers showed very good antibacterial activities. The antibacterial activity against *S. aureus* increased with the increase in alkyl chain length on ammonium group; whereas the antibacterial activity against *E. coli* decreased with increasing alkyl chain length (shown in Table II). The increase in activity against *S. aureus* is due to the fact that the polymer side chain groups with longer alkyl chains are more compatible with the bilayer structure of the bacterial cell wall. As a result of the increased compatibility, the polymer can more easily be diffused through the bacterial cell wall, rupture the cytoplasmic membrane, and kill the bacteria. The decrease in activity against *E. coli* might be due to the fact that the activity

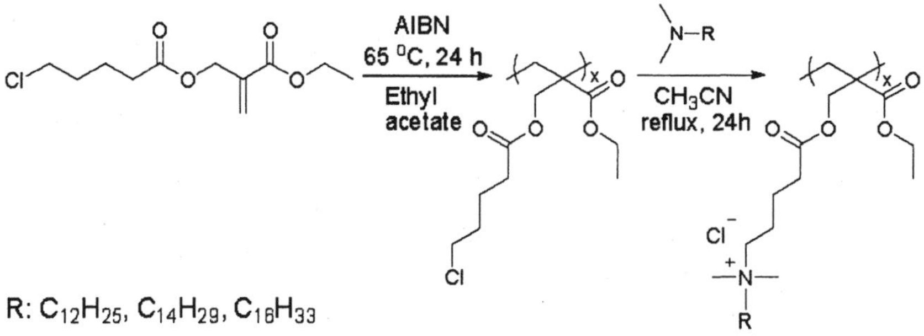

R: $C_{12}H_{25}$, $C_{14}H_{29}$, $C_{16}H_{33}$

Figure 5. Synthesis of the functional polymer and the polymers with QA groups

Table II. MBC results of the water-soluble polymers with pendant mono-QACs

Polymer	MBC (µg/mL)	
	S. aureus	*E. coli*
PCEMA-C12	128	64
PCEMA-C14	64	256
PCEMA-C16	32	256

of the polymers against *E. coli* is more dependent on the solubility of the polymer in water-TSB mixture compared to their activity against *S. aureus*.

The MBC values of the polymers at different concentrations were determined by observing the growth of bacteria on agar plates (MBC results of PCEMA-C16 shown in Figure 6). The minimum concentrations where no growth is observed are considered as the MBC values.

Water-Insoluble Methacrylate and (Meth)Acrylamide Polymers With Pendant QACs

New methacrylate and (meth)acrylamide monomers with QACs were synthesized (structures shown in Figure 7) (*84*). The first series of monomers were the derivatives of 3-(acryloyloxy)-2-hydroxypropyl methacrylate (AHM) with QACs. Hydroxylated secondary and tertiary amine derivatives of AHM

*Figure 6. The MBC results of the PCEMA-C16 polymer
against S. aureus (a) and E. coli (b)*

were prepared by reacting AHM with various amines, which was followed by the quaternization with bromohexane and bromooctane to get the monomers with QACs. A second series quaternary ammonium monomers were based on acrylamide and methacrylamide derivatives. New acrylamide/methacrylamide monomers with tertiary amine groups were synthesized and quaternized with various alkyl halides. All monomers were homopolymerized and copolymerized with 2-hydroxyethylmethacrylate (HEMA).

The synthesized monomers and polymers were tested for antibacterial activities against *S. aureus* and *E. coli* bacteria by using the broth dilution method. The monomers 2a and 6a showed the best antibacterial activities against both types of bacteria among all the monomers. These two monomers demonstrated better activities against *S. aureus* than *E. coli* (Table III). It was also found that the antibacterial activity of the quaternized methacrylamide monomers (4a, 5a, 6a) increased as the alkyl chain length on nitrogen increased.

The antibacterial activity tests for the water-insoluble homopolymers and the crosslinked copolymers were carried out by mixing a certain amount of each polymer (2 mg) with a certain amount of bacteria (5×10^5 CFU/mL) in 96-well microtiter plates and observing the growth of bacteria in each well after an incubation period. The growth of both bacteria was not observed in the case of using both the homopolymers and copolymers of monomer 2a and 6a. The growth of *S. aureus* was not also observed when the homopolymer of monomer 5a was used. On the other hand, all the other homopolymers and copolymers were not active against both types of bacteria. It can be concluded from these results that the best results were obtained with the monomer 2a and 6a and their homo- and copolymers.

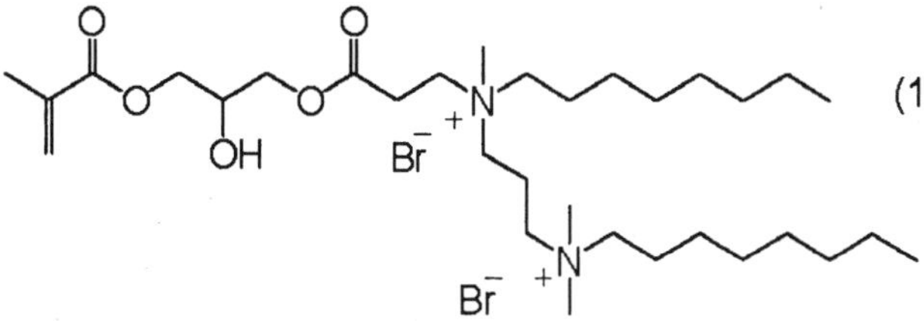

(1a)

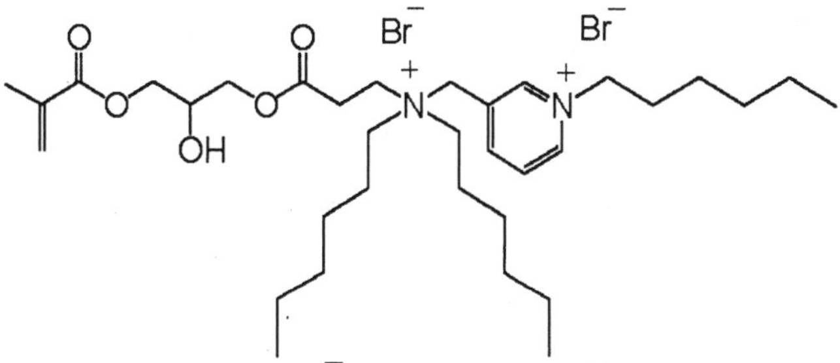

(2a)

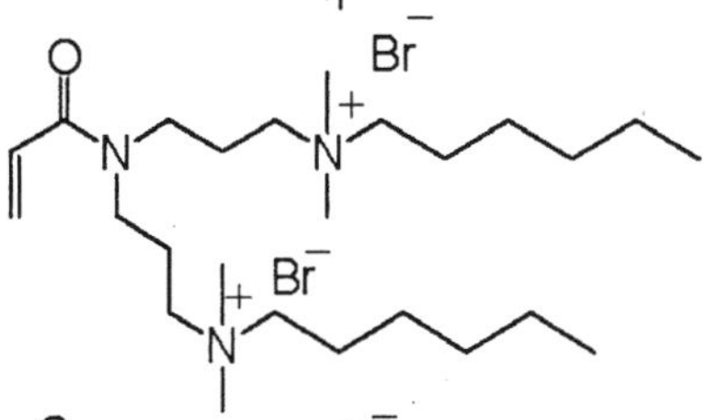

(3a)

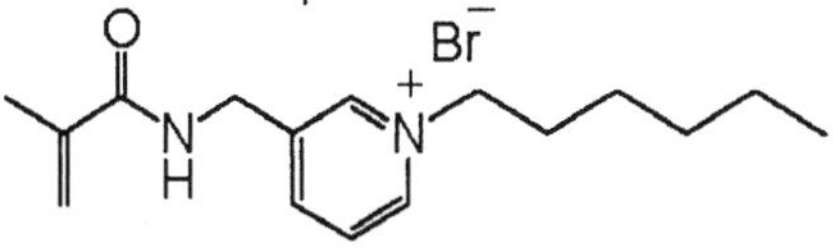

(4a)

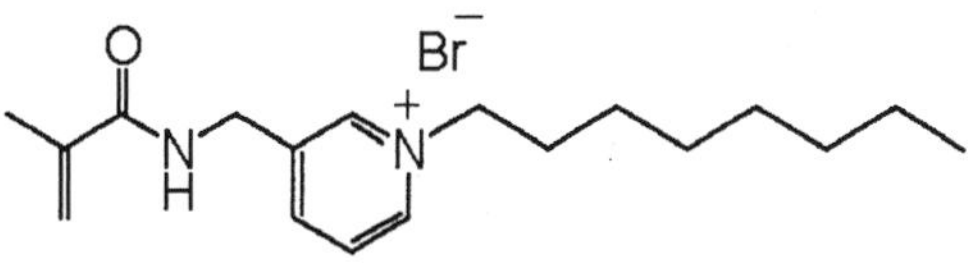

(5a)

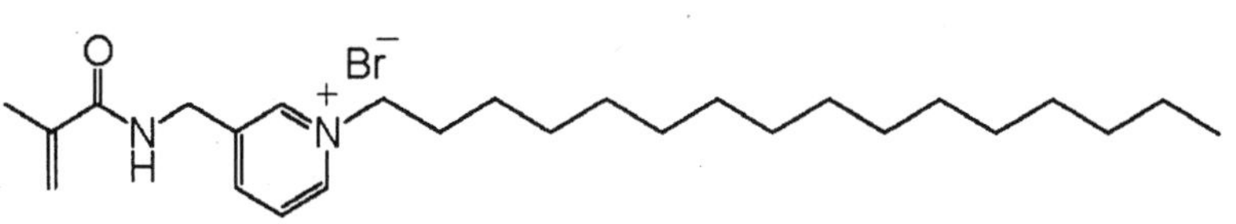

(6a)

Figure 7. The structures of the methacrylate and (meth)acrylamide monomers

Table III. Antibacterial activity results of the monomers

Monomers	MIC (μg/mL)	
	S. aureus	*E. coli*
1a	512	512
2a	32	64
3a	>512	>512
4a	>512	>512
5a	512	512
6a	2	8

Water-Insoluble Methacrylate Polymers With Pendant Norfloxacin Groups

A novel methacrylate monomer containing a norfloxacin moiety was synthesized and homopolymerized in N, N-dimethylformamide (DMF) by using azobisisobutyronitrile (AIBN) as an initiator. The new monomer was copolymerized in a 1:3 molar ratio with poly(ethylene glycol) methyl ether methacrylate (MPEGMA) in DMF using the same initiator (homopolymer and copolymer synthesis shown in Figure 8). The monomer and polymers were characterized by elemental analysis, TGA, DSC, FTIR, and NMR (*85*).

The antibacterial activities of the monomer as well as polymers were investigated against *S. aureus* and *E. coli* by using the shaking flask method, where 25 mg/mL concentrations of each compound were tested against 10^5 CFU/mL bacteria solutions. The number of viable bacteria was calculated by using the spread plate method, where 100 μL of the incubated antibacterial agent in bacteria solutions were spread on agar plates and the number of viable bacteria was counted after 24 h of incubation period at 37 °C. The monomer and polymers tested in this work showed excellent antimicrobial activities (Table IV). Although the monomer and homopolymer showed better antimicrobial activities against *E. coli* than *S. aureus*, the difference was not significant. The copolymer showed better antimicrobial activity against *S. aureus* compared to the monomer and homopolymer. Even though there is not a considerable difference, the better activity of the copolymer is probably due to the partial solubility of the copolymer in water.

Temperature Responsive (Meth)acrylamide Polymers

A new methacrylamide monomer (MAMP) containing a pyridine moiety was synthesized by reacting methacrylic anhydride and 3-(aminomethyl) pyridine. The monomer was homopolymerized in 1, 4-dioxane and copolymerized with N-isopropyl acrylamide in DMF at two different

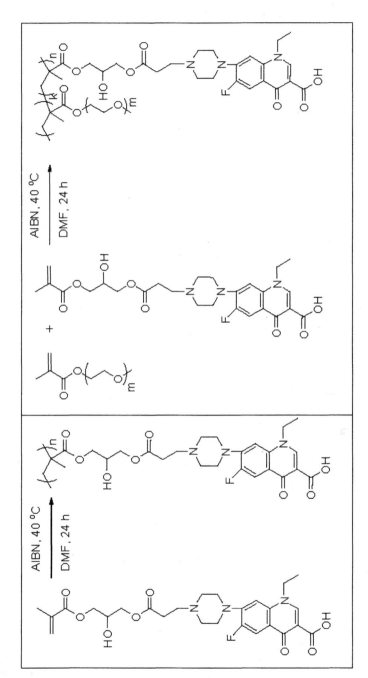

Figure 8. The syntheses of the homopolymer (left) and copolymer (right)

Table IV. Antibacterial activity results measured by Shaking Flask Method

Sample	Bacteria			
	S. aureus		*E. coli*	
	Number of Viable Bacteria	% Reduction	Number of Viable Bacteria	% Reduction
Blank	3.0×10^9	-	2.9×10^9	-
Monomer	660	100	10	100
Homopolymer	620	100	0	100
Copolymer	20	100	20	100

compositions using AIBN as an initiator. The pyridine groups of the homopolymer and copolymers were reacted with various bromoalkanes containing 12, 14, and 16 carbon alkyl chains to obtain the polymers with pendant pyridinium groups (Figure 9) (*86*). The monomer and polymers were characterized by elemental analysis, NMR, FTIR, SEC, TGA, and DSC.

The neutral and quaternized copolymers with low MAMP content were water-soluble and showed temperature responsive behavior in aqueous solutions. The lower critical solution temperatures (LCSTs) of these polymers varied between the temperatures of 25 °C and 42 °C (Figure 10). The LCST of quaternized copolymers were higher than that of the neutral copolymer since they were more hydrophilic. The LCST of the quaternized copolymers decreased with an increase in the alkyl chain length on the pyridinium group since the copolymers became more hydrophobic this way.

The antibacterial activities of water-soluble copolymers were investigated against *S. aureus* and *E. coli* using the broth dilution and spread plate methods by preparing a range of concentrations of each copolymer in water. The addition of TSB into 96-well microtiter plate wells containing different concentrations of the copolymers in water resulted in precipitation of the copolymers, which might limit the antibacterial activity. The best antibacterial activity was obtained for the 90/10 copolymer-C14 against both *S. aureus* and *E. coli*. The other quaternized water-soluble copolymers had higher MBC values; however they also showed very good activities (Table V). Considering that only around 10% of the composition of the copolymers carries pyridinium groups, these copolymers are very active. The water-insoluble polymers were also tested for the antibacterial activity against the same types of bacteria using shaking flask method. The neutral polymers and quaternized water-insoluble homopolymers and copolymers were not active.

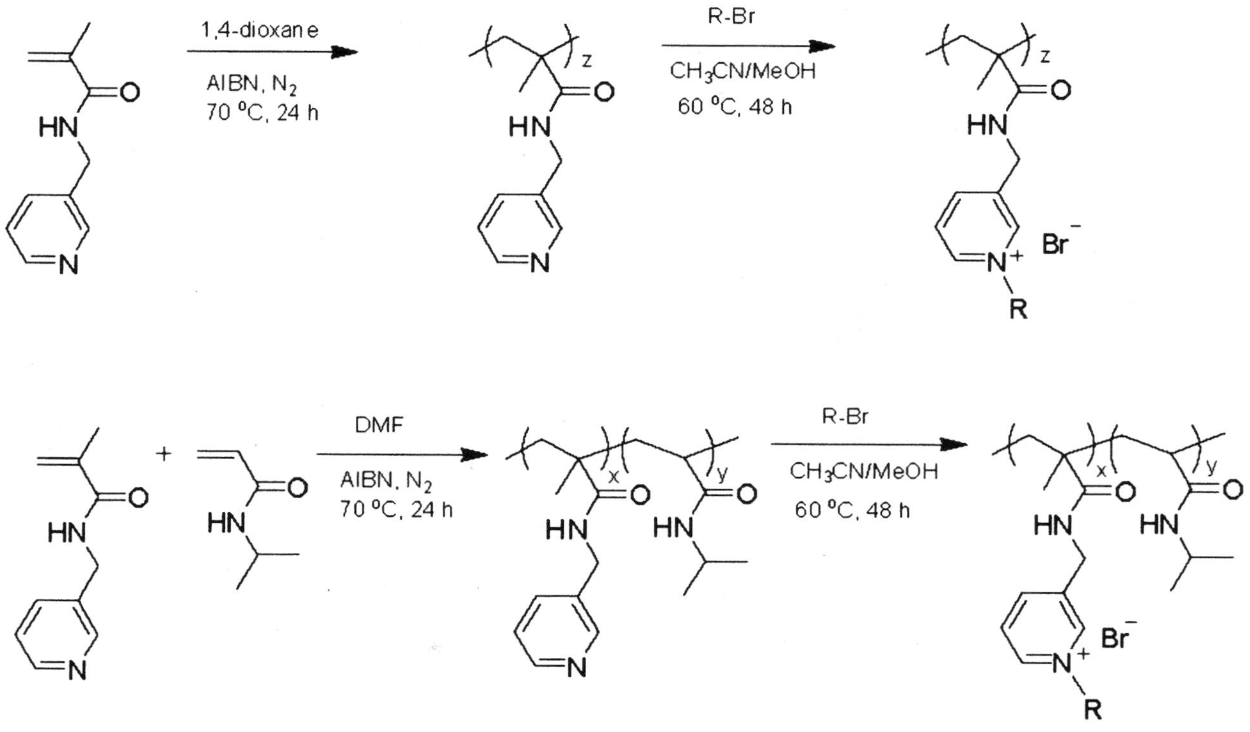

Figure 9. The syntheses of the homopolymer and copolymers with pendant pyridinium groups

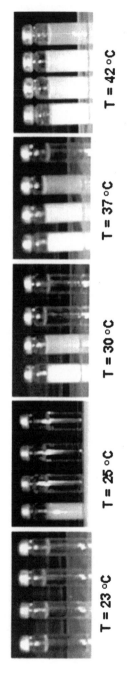

Figure 10. Phase transition temperatures for 90/10 copolymers (left to right in each photo: neat copolymer, copolymer-C16, copolymer-C14, copolymer-C12)

Table V. MBCs of the quaternized water-soluble 90/10 copolymers

Polymer	MBC (µg/mL)	
	S. aureus	*E. coli*
90/10 Copolymer-C12	640	320
90/10 Copolymer-C14	320	160
90/10 Copolymer-C16	640	320

The mechanism of action of the new antibacterial polymers with various pendant QACs against bacteria is expected to be very similar to that of existing polymeric quaternary ammonium compounds. Briefly, the polymers with pendant QACs are adsorbed onto the negatively charged cell surface by electronic interaction, and then the long lipophilic chains promote diffusion through the cell wall. The long alkyl chains disrupt the cytoplasmic membrane and cause the loss of cytoplasmic constituents, which results in the death of the microorganisms (*83,84,86*). The mode of action of the polymers with pendant norfloxacin group is due to the ability of the norfloxacin moiety in the polymers to inhibit DNA synthesis through prevention of DNA replication process (*48,57*). In order to be active, norfloxacin moiety should be cleaved from the polymers and present inside the cytoplasmic membrane. There are a few possibilities for the cleavage of norfloxacin moieties from the polymers. The first possibility is that the two ester bonds in the polymers can hydrolyze to give norfloxacin or a norfloxacin derivative. The second possibility is that the Michael addition can be reversed to release norfloxacin from the polymers (*85*).

Conclusions

In conclusion, a detailed literature information was provided about low molecular weight antibacterial agents (QACs and norfloxacin), antibacterial polymers with QACs, and antibacterial activity tests in the first part of the review. In the second part, the syntheses and antibacterial activities of several new antibacterial polymers with pendant QACs and norfloxacin groups against *S. aureus* and *E. coli* were discussed. The polymers were structurally designed and synthesized to demonstrate excellent antibacterial activities against both types of bacteria. Effective structural parameters were determined and adjusted to optimize the antibacterial activity. In general, all polymers showed very good antibacterial activities against both types of bacteria. Some of the trends obtained by studying these polymers can be summarized as follows. First, the polymers showed better antibacterial activities than the corresponding monomers. Second, the antibacterial activity of the polymers with QACs increased as the chain length attached to the QA group increased. Third, the water-soluble polymers showed better antibacterial activities than water-insoluble polymers due to the fact that they interact better with the bacteria. Fourth, the polymers demonstrated better antibacterial activities against *S. aureus* than *E. coli*, which is a reasonable result since *E. coli* is a Gram-negative bacteria and harder to kill than *S. aureus*, which a Gram-positive bacteria. These new antibacterial polymers can be used in a number of applications including but not limited to coatings, water treatment, contact lenses, biomedical and pharmaceutical applications.

References

1. Ikeda, T.; Tazuke, S.; Suzuki, Y. *Makromol. Chem.* **1984**, *185*, 869.
2. Mcdonnell, G.; Russell, A. D. *Clin. Microbiol. Rev.* **1999**, *12*(1), 147.
3. Ranucci, E.; Ferruti, P.; Neri, M. G. *J. Biomater. Sci. Polym. Ed.* **1991**, *2*(4), 255.
4. Granger, R.; Koeberle, J.; Hao-Dong, L.; Yavordios, D. *Chimica Therapeutica* **1968**, *3*(2), 129.
5. Tashiro, T. *Macromol. Mat. Engin.* **2001**, *286*, 63.
6. Gabrielska, J.; Sarapuk, J.; Przestalski, S.; Wroclaw, P. *Tenside, Surfactants, Detergents*, **1994**, *31*(5), 296.
7. Przestalski, S.; Sarapuk, J.; Kleszczynska, H.; Gabrielska, J.; Hladyszowski, J.; Trela, Z.; Kuczera, J. *Acta Biochimica Polonica*, **2000**, *47*(3), 627.
8. Sauvet, G.; Dupont, S.; Kazmierski, K.; Chojnowski, J. *J. Appl. Polym. Sci.* **2000**, *75*, 1005.
9. Merianos, J. J. In *Block, S. S. eds.* Disinfection, Sterilization, and Preservation, 4th edn. Pennsylvania: Lea & Febiger **1991**, pp. 225.
10. Talaro, K.; Talaro, A. In *Foundations in Microbiology*, WCB Publishers **1993**, 286.
11. Li, G.; Shen, J.; Zhu, Y. *J. Appl. Poly. Sci.* **2000**, *78*, 668.
12. Russell, A. D. Quaternary Ammonium Compounds www.maunco.com/seminars/transcripts/QACs-1.ppt, Sept. 30, **2005**.
13. Brown, G. E. Postharvest Florida Citrus Guide. www.fred.ifas.ufl.edu/citrus/srd.pdf, September 30, **2005**.
14. Olejnik, O; Kerslake, E. D. S. US Patent **2003**, 6562873.
15. Tan, S.; Li, G.; Shen, J.; Liu, Y., Zong, M. *J. Appl. Polym. Sci.* **2000**, *77*(9), 1869.
16. Li, G.; Shen, J. R.; Zhu, Y. *J. Appl. Polym. Sci.* **1998**, *67*, 1761.
17. Ikeda, T.; Hirayama, H.; Yamaguchi, H.; Tazuke, S.; Watanabe, M. *Antimicrob. Agents Chemother.* **1986**, *30*(1), 132.

18. Ikeda, T. *High Perform. Biomater.* **1991**, 743.
19. Kenawy, E.; Abdel-Hay, F. I.; El-Raheem, A.; El-Shanshoury, R.; El-Newehy, M. H. *J. Polym. Sci: Part A: Polym. Chem.* **2002**, *40*, 2384.
20. Hazziza-Laskar, J.; Helary, G.; Sauvet, G. *J. Appl. Polym. Sci.* **1995**, *58*, 77.
21. Ikeda, T.; Yamaguchi, H.; Tazuke, S. *Antimicrob. Agents Chemother.* **1984**, *26*(2), 139.
22. Dizman, B.; Elasri, M. O.; Mathias, L. J. *J. Appl. Polym. Sci.* **2004**, *94*(2), 635.
23. Birnie, C. R.; Malamud, D.; Schanaare, R. L. *Antimicrob. Agents Chemother.* **2000**, *44*, 2514.
24. Abel, T.; Cohen, J. I.; Engel, R.; Filshtinskaya, M.; Melkonian, A.; Melkonian, K. *Carbohydr. Res.* **2002**, *337*(24), 2495.
25. Cohen, J. I.; Abel, T.; Filshtinskaya, M.; Melkonian, K.; Melkonian, A.; Burkett, D.; Engel, R. *Abstracts of Papers, 223rd ACS National Meeting, Orlando, FL, United States* **2002**, CARB-059.
26. Borman, S. *Sci. Tech.* **2001**, *79*(22), 13.
27. Tiller, J. C.; Liao, C. J.; Lewis, K.; Klibanov, A. M. *Proc. Nati. Acad. Sci.* **2001**, *98*(11), 5981.
28. Nurdin, N.; Helary, G.; Sauvet, G. *J. Appl. Polym. Sci.* **1993**, *50*, 663.
29. Nurdin, N.; Helary, G.; Sauvet, G. *J. Appl. Polym. Sci.* **1993**, *50*, 671.
30. Kawabata, N.; Nishiguchi, M. *Appl. Environ. Microbiol.* **1988**, *54*(10), 2532.
31. McKenna, S. Long Island University. www.liu.edu/cwis/bklyn/acadres/facdev/FacultyProjects/WebClass/micro-web/html-files/chapterA1.html, Jan. 10, **1997**.
32. Kanazawa, A.; Ikeda, T.; Endo, T. *J. Polym. Sci: Part A: Polym. Chem.* **1993**, *31*, 335.
33. Kenawy, E. *J. Appl Poly. Sci.* **2001**, *82*, 1364.
34. Beveridge, T. J. *J. Bacteriol.* **1999**, *181*, 4725.
35. Schroeder, J. D.; Scales, J. C. US Patent **2002**, 20020051754.
36. Rosinskaya, C.; Weinberg, A. US Patent **2004**, 20040106912.
37. Tebbs, S. E.; Elliott, T. S. J. *J. Antimicrob. Chemother.* **1993**, *31*(2), 261.
38. Kawabata, N.; Fujita, I.; Inoue, T. *J. Appl. Polym. Sci.* **1996**, *60*, 911.
39. Broughton, R. M.; Worley, S. D.; Slaten, B. L.; Mills, G.; Sunderman, C.; Sun, G.; Michielsen, S. In *National Textile Center Annual Report* **1998**, 347.
40. Sun, G. International Patent **2000**, WO 00/15897.
41. Kyba, E. P.; Park, J. US Patent **2000**, 6051611.
42. Vigh, J. E.; Lo, P.; Dziabo, A. J.; Wong, M. P. US Patent **1994**, 5277901.
43. Robertson, J. R. US Patent **1994**, 5358688.
44. Fang, K.; Chen, Y.; Sheu, J.; Wang, T.; Tzeng, C. *J. Med. Chem.* **2000**, *43*, 3809.
45. Aeschlimann, J. R.; Dresser, L. D.; Kaatz, G. W.; Rybak, M. *J. Antimicrob. Agents Chemother.* **1999**, *43*(2), 335.

46. Hooper, D. C.; Wolfson, J. S. *Quinolone Antimicrobial Agents*, 2nd ed.; American Society of Microbiology: Washington, DC, **1993**, 53.

47. Parshikov, I. A.; Freeman, J. P.; Lay, J. J. O.; Beger, R. D.; Williams, J.; Sutherland, J. B. *Appl. Environ. Microbiol.* **2000**, *66*(6), 2664.

48. Golet, E. M.; Alder, A. C.; Hartmann, A.; Ternes, T. A.; Giger, W. *Anal. Chem.* **2001**, *73*, 3632.

49. Koga, H.; Itoh A.; Murayama, S.; Suzue, S.; Irikura, T. *J. Med. Chem.* **1980**, *23*(12), 1358.

50. Kondo, H.; Sakamoto, F.; Inoue, Y.; Tsukamato, G. *J. Med. Chem.* **1989**, *32*(3), 679.

51. Ledoussal, B.; Bouzard, D.; Coroneos, E. *J. Med. Chem.* **1992**, *35*(1), 198.

52. Chen, Y.; Fang, K.; Sheu, J.; Hsu, S.; Tzeng, C. *J. Med. Chem.* **2001**, *44*, 2374.

53. Oliphant, C. M.; Green, G. M. *Am. Fam. Physician* **2002**, *65*(3), 455.

54. Chu, D. T. W.; Fernandes, P. B.; Claiborne, A. K.; Pihuleac, E.; Nordeen, C. W.; Maleczka, R. E.; Pernet A. G. *J. Med. Chem.* **1985**, *28*, 1558.

55. Chu, D. T. W.; Fernandes, P. B.; Pernet A. G. *J. Med. Chem.* **1986**, *29*, 1531.

56. Domagala, J. M.; Heifetz, C. L.; Hutt, M. P.; Mich, T. F.; Nichols, J, B.; Solomon, M.; Worth, D. F. *J. Med. Chem.* **1988**, *31*, 991.

57. Hwangbo, H. J.; Lee, Y.; Park, J. H.; Lee, Y. R.; Kim, J. M.; Yi, S.; Kim, S. K. *Bull. Korean Chem. Soc.* **2003**, *24*(5), 579.

58. Trooskin, S. Z.; Dontetz, A. P.; Harvey, R. A.; Greco, R. S. *Surgery* **1985**, *97*, 547.

59. Modak, S. M.; Sampath, L.; Fox, C. L.; Benvenisty, A.; Nowygrod, R.; Reemstmau, K. *Surg. Gynecol. Obstert.* **1987**, *164*, 143.

60. Bach, A,; Schmidt, H.; Bottiger, B.; Schreiber, B.; Bohrer, H.; Motsch, J.; Martin, E.; Sonntag, H. G. *J. Antimicrob. Chemother.* **1996**, *37*, 315.

61. Phaneuf, M. D.; Ozaki, C. K.; Bide, M. J.; Quist, W. C.; Alessi, J. M.; Tannenbaum, G. A.; Logerfo, F. W. *J. Biomed. Mater. Res.* **1993**, *27*, 233.

62. Schierholz, J. M.; Rump, A.; Pulverer, G. *Arzneim.-Forsch.* **1997**, *47*, 70.

63. Lowman, A. M.; Peppas, N. A. *Macromolecules* **1997**, *30*, 4959.

64. Roseeuw, E.; Coessens, V.; Schacht, E.; Vrooman, B.; Domurado, D.; Marchal, G. *J. Mater. Sci.: Mater. Med.* **1999**, *10*, 743.

65. Coessens, V.; Schacht, E.; Domurado, D. *J. Controlled Release* **1997**, *47*, 283.

66. Yang, M.; Santerre, J. P. *Biomacromolecules* **2001**, *2*, 134.

67. Moon, W.; Kim, J. C.; Chung, K.; Park, E.; Kim, M.; Yoom, J. *J. Appl. Polym. Sci.* **2003**, *90*, 1797.

68. Gil, E. S.; Hudson, S. M. *Prog. Polym. Sci.* **2004**, *29*, 1173.

69. Pan, Y. V.; Wesley, R. A.; Luginbuhl, R.; Denton, D. D.; Ratner, B. D. *Biomacromolecules* **2001**, *2*, 32.

70. Kuckling, D.; Harmon, M. E.; Frank, C. W. *Macromolecules* **2002**, *35*, 6377.

71. Cho, E. C.; Lee, J.; Cho, K. *Macromolecules* **2003**, *36*, 9929.

72. Motokawa, R.; Morishita, K.; Koizumi, S.; Nakahira, T.; Annaka, M. *Macromolecules* **2005**, *38*, 5748.
73. Brazel, C. S.; Peppas, N. A. *Macromolecules* **1995**, *28*, 8016.
74. Stile, R. A.; Healy, K. E. *Biomacromolecules* **2001**, *2*, 185.
75. Kim, S.; Healy, K. E. *Biomacromolecules* **2003**, *4*, 1214.
76. Schilli, C. M.; Zhang, M.; Rizzardo, E.; Thang, S. H.; Chong, Y. K.; Edwards, K.; Karlsson, G.; Muller, A. H. E. *Macromolecules* **2004**, *37*, 7861.
77. Zhang, W.; Shi, L.; Wu, K.; An, Y. *Macromolecules* **2005**, *38*, 5743.
78. Rueda, J.; Zschoche, S.; Komber, H.; Schmaljohann, D.; Voit, B. *Macromolecules* **2005**, *38*, 7330.
79. White, D. G.; Acar, J.; Anthony, F.; Franklin, A.; Gupta, R.; Nicholls, T.; Tamura, Y.; Thompson, S.; Threlfall, E. J.; Vose, D.; Vuuren, M. V.; Wegener, H. C.; Costarrica, M. L. *Rev. Sci. Tech. Off. Int. Epiz.* **2001**, *20*(3), 849.
80. Goodson, B. A.; Ehrhardt,A.; Ng, S.; Nuss, J.; Johnson, K.; Giedlin, M.; Yamamoto, R.; Moos, W. H.; Krebber, A.; Ladner, M.; Giacona, M. B.; Vitt, C.; Winter, J. *Antimicrob. Agents Chemother.* **1999**, *43*, 1429.
81. Moon, W.; Kim, J. C.; Chung, K.; Park, E.; Kim, M.; Yoom, J. *J. Appl. Polym. Sci.* **2003**, *90*, 1797.
82. Dizman, B.; Mathias, L. J. *J. Polym. Sci. Part A: Polym. Chem.* **2005**, *43*, 5844.
83. Dizman, B.; Elasri, M. O.; Mathias, L. J. *J. Polym. Sci. Part A: Polym. Chem.* **2006**, *44*(20), 5965.
84. Ayfer, B.; Dizman, B.; Elasri, M. O.; Mathias, L. J.; Avci, D. *Designed Monomers and Polymers* **2005**, *8*, 437.
85. Dizman, B.; Elasri, M. O.; Mathias, L. J. *Biomacromolecules* **2005**, *6*, 514.
86. Dizman, B.; Elasri, M. O.; Mathias, L. J. *Macromolecules* **2006**, *39*(17), 5738.

Chapter 3

Engineering Nanoporous Bioactive Smart Coatings Containing Microorganisms: Fundamentals and Emerging Applications

**M. C. Flickinger[1,2], M. Fidaleo[1], J. Gosse[1,2], K. Polzin[1],
S. Charaniya[1,5], C. Solheid[1], O. K. Lyngberg[1,5], M. Laudon[1], H. Ge[5],
J. L. Schottel[2], D. R. Bond[3], A. Aksan[4], and L. E. Scriven[5]**

[1]BioTechnology Institute, Departments of [2]Biochemistry, Molecular Biology
and Biophysics, [3]Microbiology, [4]Mechanical Engineering, and [5]Chemical
Engineering and Materials Science, University of Minnesota,
Minneapolis and Saint Paul, MN 55455

Nanoporous, adhesive latex coatings and ink-jet deposited latex microstructures containing concentrated, viable, but nongrowing microorganisms may be useful smart coatings. When rehydrated, these bioactive coatings can be used for multi-step oxidations, reductions, as biosensors, in biofuel cells, or high intensity industrial biocatalysts. Engineering coating microstructure, preservation of microbe viability during drying at ambient temperature and the stability of these coatings following rehydration is investigated in 5 μm to 75 μm thick coatings of microbes concentrated 10^2 to 10^3-fold on polyester, metals or electrode substrates. Nanoporosity is essential for preserving microbial viability in dry coatings and bioreactivity following rehydration. Non-toxic (low biocide or biocide-free) latex emulsions contain carbohydrate porogens which vitrify to arrest polymer particle coalescence during film formation generating nanopores. However, the molecular mechanism of how vitrified carbohydrates function as osmoprotectants and preserve microbial viability by formation of glasses in the pore space during film formation is unknown. Coating nanoporosity in hydrated films is estimated by tracer diffusivity and visualized by cryogenic-SEM. Emulsion composition, drying conditions and coating thickness affect microbial viability, substrate adhesion, and coating reactivity following drying, storage and rehydration. The specific reactivity of the entrapped microorganisms can be induced to express enzymes for optimal reactivity prior to coating or the microbes can be "activated" by inducing gene expression following coat drying and rehydration. Laser scanning confocal microscopy is used to investigate spatial gene expression as a function of coating depth and diffusion resistance. Model microbial smart coatings investigated include: an *E. coli* ionic mercury biosensor, an anaerobic starch hydrolyzing coating of *Thermotoga maritima* at 80°C, photoreactive coatings of *Rhodopseudomonas palustris* for anoxic production of hydrogen, coatings of *Gluconobacter oxydans* which oxidizes D-sorbitol → L-sorbose, and current-generating coatings of *Geobactor sulfurreducens* on conductive electrode materials.

Biocatalytic coatings are an emerging class of smart coatings (*1, 2*). Biocatalytic coatings include water-borne coatings engineered to stabilize, concentrate, and intensify the reactivity of enzymes or cells entrapped in a nanostructured adhesive coating formed by a polymer matrix or partially coalesced polymer particles.

Biocatalytic coatings containing entrapped or chemically bound enzymes have been used as reactive smart sensors to detect environmental contaminants, as self-cleaning surfaces, textile coatings, or to catalytically decontaminate surfaces from exposure to chemical warfare agents (*2*). While enzyme stability and activity can be manipulated by well developed protein engineering methods and enzymes can be efficiently manufactured by microbial recombinant DNA technology, smart coatings containing enzyme–based additives are capable of only single-step hydrolytic reactions (*2-9*); only recently have nanoporous materials containing more than one immobilized enzyme been reported for multi-step biocatalysis (*10*). An additional limitation of enzyme-containing smart coatings is that many isolated enzymes cannot carryout oxidizations or reductions because they require expensive cofactors (such as ATP for phosphorylation or release of energy by hydrolysis) and biological electron donors (reduced co-enzymes NADH, FADH, or reduced ferredoxin). While *in vitro* biological cofactor regeneration systems including cofactor "tethering" have been reported for immobilized enzyme systems (*10*), it is more cost effective and robust to use living cells to regenerate cofactors and therefore industry still uses slowly growing, entrapped, or "resting" microbes for selective oxidations and reductions to produce chemical intermediates, for environmental decontamination and waste treatment.

Why Develop Smart Coatings of Nongrowing Microorganisms Preserved at Ambient Temperature?

Much effort has been expended in the coating industry to eliminate or inhibit the growth of microorganisms in latex emulsions in order to extend paint shelf-life and functionality following coating and film formation. Microbial smart coatings are the opposite - adhesive latex emulsions that are non-toxic to a very high concentration of microbes in order to preserve microbial viability following film formation. Why would living microbes be useful reactive additives in smart coatings?

Living microbes are highly selective catalysts capable of a vast array of multi-step chiral reactions (*11, 12*) and surprisingly many of these reactions can be carried out by nongrowing cells. Microorganisms can detoxify environmental pollutants, produce a variety of useful products (peptides, proteins, alcohols, organic acids, biopolymers, pharmaceutically active compounds such as antibiotics, or useful gases such as H_2 or methane), or transport electrons to an electrode surface. Extremeophile microorganisms are capable of carrying out these reactions in harsh chemical environments (extreme pH, salinity, pressure, temperatures >100°C) and therefore are attractive "green" biocatalysts. Microbes can be genetically manipulated using recombinant DNA technology to alter their capabilities as biocatalysts. However, living cells as biocatalysts are seldom available as "off-the-shelf" (stabilized) highly reactive (concentrated) reagents. Most microbes cannot be stored in a partially desiccated state without loss of activity unless frozen, are used in dilute suspension, and have insufficient stability (active half-life) and specific reactivity (intensity) to replace existing chemical catalysts in industrial processes.

In contrast to dilute suspensions, most chemical catalysis is carried out in the pore space of robust catalytic media. The rate of biocatalysis could be significantly increased if living microorganisms also could be immobilized at high density on surfaces with minimal diffusion resistance in coatings that remain adhesive when continuously irrigated with water. Using coating technology to preserve non-growing living bacteria in thin (<50 μm) nanoporous adhesive latex coatings could provide an inexpensive method to stabilize their viability and activity and increase their volumetric reactivity (intensity) similar to that of chemical catalysts.

Engineering non-toxic nanoporous latex coating formulations for preservation of the viability and reactivity of a high density of living microbes without refrigeration following film formation and partial dehydration (coat drying) is a new and challenging problem for the coating industry. Solving this problem will significantly expand the use of microbes as biosensors, as bio-protective or anticorrosive coatings, bioactive architectural coatings, components of microbial fuel cells and could dramatically expand the use of biocatalysts. It will also expand the use of water-borne multi-layer coating technologies for generating large scale bioreactive surfaces, as well as expand the use of ink-jet latex emulsion printing methods to generate miniature coatings for bionano or bioelectronic device applications (*1, 13*). In addition, by using systems biology, metabolic and protein engineering methods, microbes could be genetically optimized for use in smart coatings to have comparable reactivity and half-lives as chemical catalysts (thousands of hours).

Development of Polymer Coatings Containing Living Microorganisms

Living cells have been entrapped in many types of modified cross-linked sol-gel synthetic polymers, in sol-gel oxide ceramics (biocers) and silica gel glasses (*14-21*). Although mechanically robust (hard, transparent, often reinforced using fibers, formed by freeze-casting methods), ceramic coatings made from silica or inorganic nanosols often kill microbes during cross-linking to form solvent-containing lyogels due to alcohol toxicity. Cell viability is further reduced during lyogel drying to form stronger xerogel films due to shrinkage of the silica network (drying at elevated temperature or freeze-drying). Alcohol toxicity can be minimized by using aqueous nanosols produced by metal-catalyzed silicon or metal alkoxide hydrolysis in water. The addition of glycerol (up to 16% by wt) (*21*) as well as the addition of pore-forming monosaccharides such as sorbitol (up to 50% by wt) are important to retain high cell viability (~70 % viable cells) and high coating porosity (50 to 60% measured by mercury intrusion). Average pore diameters measured by BET are 7 nm ~10 nm (*20*). However, the major disadvantage of sol-gel ceramics for biologically reactive coatings is that only 5% to 20% by wt of cell mass can be incorporated. Higher concentrations of living cells reduce the stability of these coatings when in contact with water. Therefore their reactivity m^{-2} is very low and their ability to retain entrapped cell viability after storage at 4°C is poor (<5% cell viability after 9 days) (*20*).

There are many publications describing hydrogel-based microbial polymer immobilization systems (gel beads, slabs or gel-coated beads) because they are easily fabricated in the laboratory and retain entrapped cells without loss of viability (*22*). Unfortunately hydrogels are not useful as coatings because they are not adhesive, have poor mechanical strength, and are 100s of microns to mms thick resulting in severe mass transfer limitations. Hydrogels are macro porous (pores larger than the microorganisms), have thin pore walls, and cannot be stored dry or frozen for long periods of time without loss of cell viability (*22, 23*). When hydrogel-entrapped microorganisms are incubated in the presence of nutrients to sustain cell viability and regenerate activity, significant release and outgrowth of viable cells occurs. Most hydrogels also require covalent cross-linking to enhance their mechanical stability. In spite of the numerous publications (*22–29*), hydrogel immobilization is not sufficiently stable, reactive, or adhesive nor capable of preserving the viability of microbes during gel desiccation.

Bioreactive water-borne latex polymer coatings containing enzymes or antibodies with color indicators were developed more than 25 years ago by Eastman Kodak for use in layered dry reagent clinical chemistry (30, 31), but these early studies did not include coatings of living cells.

Microbial latex biocatalytic coatings originated with investigations by Lawton, Bunning and Flanagan in the 1980s that first used polymer laticies to coat solid particles, nylon mesh, membranes, and silica (33-36). These studies used a polydispersed acrylate/vinyl acetate copolymer (Polyco 2151, ave. particle size 260 nm, T_g = 13°C, Rohm & Haas). Coating porosity was generated by integrating calcium carbonate which was later leached from the coatings with acid to allow pores colonized by microbial growth. Cantwell (37) first reported the use of polymer blends of hard and soft polymer particles having a T_g range of -60°C up to 60°C for microbial entrapment to stabilize enzyme activity. However, this study did not coat these blends in thin coatings – only flocculates, 1-2 mm diameter aggregates and 2 mm diameter fibrils. No data was presented on cell viability following entrapment or aggregate permeability. Martens and Hall (38) reported methacrylate and acrylate polymer coatings of *Synechococcus* on a carbon electrode using a variety of acrylate polymers. Upon rehydration and exposure to light, photosynthetic activity was measured in an electrochemical cell. Coating diffusion properties were measured using a rotating disc electrode. Cell viability was reported to be "nearly 100%". The major difficulties of these early polymer coating investigations for entrapment of living cells were: low coating permeability, coating instability (delaminating from the support particles), poor control of coating thickness, no methods to accurately measure the number of viable cells that survived the latex coating/drying/rehydration process, and lack of fundamental knowledge of film formation and how these polymer coatings preserved microbial viability during coat drying at ambient temperature. In none of these early investigations was the nanoporosity, pore microstructure, reactivity (intensity, effectiveness factor), uniformity of the coatings (single or multi-layer), or the viability measured nor was the physiology (response to drying stress) of the latex-embedded microorganisms investigated.

Swope and Flickinger were the first investigators to demonstrate non-growth gene expression and β-galactosidase enzyme synthesis in 80 μm thick bilayer acrylate/vinyl acetate latex *E. coli* coatings containing glycerol to preserve viability at ambient temperature under nitrogen-limited conditions (39). *E. coli* retained >90% viability following coating onto polyester, drying, storage, and rehydration for several weeks, and were retained by a nanoporous latex top coat (40). Differential viability staining and laser scanning confocal microscopy (LSCM) with image analysis were used to determine that the viability of the entrapped *E. coli* was as high as 95% for several weeks, however, this method was limited to penetration into the coatings of only 5 μm due to photobleaching (39 - 42). Subsequent work by Lyngberg using 30 μm thick *E. coli* coatings with a nanoporous sealant top coat showed that the addition of sucrose or trehalose with glycerol could increase coating porosity

(43, 44). A circular patch coating method on polyester was developed using perforated pressure-sensitive vinyl mask templates and Mayer rod draw down coating. The nanoporous microstructure of these coatings was investigated using cryogenic scanning electron microscopy (cryo-SEM) (43-45, 47-49) and a single-use patch biosensor was developed to detect ionic mercury (Hg^{+2}) (46). A bimodal blend polymer formulation of acrylate and polystyrene was also developed for the hyperthermophilic marine bacterium *Thermotoga maritima* to devise a biocatalytic coating with stable pores at 80°C to concentrate and stabilize the toga-associated amylase activity of this salt-tolerant anaerobe (50).

Engineering Coating Reactivity

Microbial smart coating reactivity (intensity) is a function of entrapped cell density, viability, thinness and nanoporosity following film formation and rehydration. Latex biocatalytic coatings cast using simple laboratory methods can be ~5 μm to ~75 μm thick and thus can overcome some of the mass transfer limitations of hydrogel entrapment in spite of their reduced porosity compared to gels. Latex coatings also maintain their adhesiveness and integrity at cell concentrations as high as 50% by volume in contrast to hydrogels which disintegrate when the cell concentration exceeds 30% (23).

The focus of investigations during the late 1980s and 1990s by the group at the University of Minnesota was on engineering coating nanoporosity by controlling the degree of polymer particle coalescence of low T_g latex emulsions by quenching using temperature, porogens, bimodal blends, or core-shell approaches in order to create diffusive coatings (Figure 1). Nanoporosity permanently entraps micron-scale microorganisms in adhesive polymer because the pores are smaller than the microbes. Nanoporosity is generated by vitrification of glycerol and carbohydrates in the pore space to arrest polymer particle coalescence. By varying polymer particle properties and porogens, water-borne formulations could be engineered to generate coatings with stable pores at elevated temperature, extremes of pH, or in high salt environments. Bimodal blend and core-shell approaches can minimize wet coalescence following coating rehydration; permanently porous latex coatings useful for arresting wet coalescence have been described (51).

In addition to nanoporosity, coatings must be adhesive (to substrate, layer to layer), contain a very high volume fraction of microorganisms below the critical concentration which significantly decreases porosity and disrupts polymer particle coalescence, and concentrate the cells at least 10^2 to 10^3-fold from the density achieved in a growing microbial culture (53). Most critical is that latex emulsions are formulated so that they are non-toxic, do not contain growth inhibiting levels of biocides or resides from solution polymerization (see below).

An alternative form of diffusive bioactive smart coatings is a perfusive stand-alone all latex tri-layer membrane with microorganisms permanently

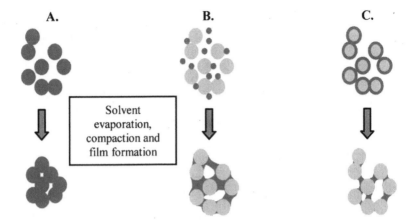

Figure 1. Nano-pore formation by arrested (quenched) coalescence of latex polymer particles. A. Monodispersed low T_g latex. B. Bimodal blend (large high T_g + small low T_g). C. Core-shell latex, high T_g core, low T_g shell. (Adapted from reference 52)

entrapped between two thin nanoporous polymer layers or emulsions suitable for adhesion to pre-formed porous membranes (*43, 44*). With efficient mass transfer, the reactivity of perfusive bioactive or biocatalytic membrane structures would be limited only by the permeability of the entrapped microbes. Jons (*54*) reported fabricating porous latex membranes; however, the methods used organic solvents which kill microorganisms. Development of organic solvent-free emulsions to cast perfusive latex membranes containing viable microorganisms that withstand a significant pressure or adhesive latex for coating pre-formed membranes would enable the concentration of biocatalysts at phase interfaces (gas/liquid, gas/solid, aqueous/non-aqueous) to engineer high intensity multi-phase membrane bioreactors.

Activating Microbial Smart Coatings

Model microbial smart coatings have been investigated under nitrogen-limited conditions to inhibit microbial out-growth. Most coatings use a nanoporous sealant polymer-only topcoat to retain the cells embedded in the layers below. Phosphate and an energy source such as pyruvate, organic acids, or light sustain cell viability without out-growth. Up-regulation of gene expression and subsequent protein synthesis by the embedded microbes is used to activate the coatings or to generate responsive or biosensing coatings. Gene induction can increase coating specific reactivity by several orders of magnitude and molecular biology methods can be used to alter response kinetics.

The demonstration of β-galactosidase expression over a period of 17 days induced by lactose or IPTG in non-growing latex-entrapped *E. coli* by Swope and Flickinger (*39*) was the first report of *in situ* activation of a microbial latex coating. Recently Desmone (*55*) reported a hydrophilic silicon oxide matrix for preservation of *E. coli* for recombinant protein production, but only as a starter culture, not to activate the coating or change its specific activity without growth.

Two approaches are possible to preserve coating reactivity during dry storage/ shipment and to significantly extend rehydrated coating activity (half-life) beyond that of suspended non-growing "resting cells" (*1*). In the first approach, microbial coatings are cast pre-induced or "pre-activated". The microbes are induced to accumulate high levels of intracellular enzymes during growth (prior to coat casting) so that they contain the desired biocatalytic reactivity. This approach is used when the enzymes for the desired reactivity are not in a soluble form in the cytoplasm of the cells but are membrane-associated (*52, 53, 56*). Because the cells can be preserved in a viable state in the coatings without growth by being partially desiccated during coat drying, membrane-associated enzyme reactivity is preserved. Following rehydration, *in vivo* enzyme activity may also be significantly stabilized by membrane association. However, as a result of normal *in vivo* protein turnover, intracellular enzyme activity will slowly decay. The potential exists to periodically induce new gene expression and renew enzyme synthesis in coating-entrapped cells by limiting additions of a source of nitrogen or growth stimulating nutrient thus "re-activating" the coating (*1*).

In the second approach, microbial biocatalytic coatings can be made, stored, and shipped to the site of use without the cells having the desired enzymes for a particular application – an "inactive" coating. Following rehydration at the site of use, expression of the genes encoding the enzymes for that particular application can be induced by the addition of a suitable inducer and a nitrogen source and the coating "activated". This approach is favored for soluble cytoplasmic or periplasmic enzymes (*39*). In this second approach, microbial biocatalytic coatings will not lose the specific enzyme activity needed for a particular application during dry storage and shipping prior to use because they are manufactured in an inactive (un-induced) state. These approaches which generate coatings of living cells preserved at ambient temperature separate the site of biocatalyst manufacture from the site of biocatalyst use.

Methods for Investigating Microbial Biocatalytic Coatings

Polymer Toxicity

A variety of acrylate, methacrylate, vinyl acetate, butadiene and polystyrene latex polymers, copolymers and polymer blends containing different biocides and residual monomers have been evaluated for toxicity to bacteria, yeast and fungi (*30-36, 38*). In our work evaluating low T_g acrylate, vinyl acetate and polystyrene copolymer latexes and blends adjusted to neutral pH immediately prior to use, 30 min. contact toxicity assays (CTA) followed by viable plate counting and a film formation toxicity assay (see below) are used to measure toxicity. The toxicity of latex emulsion components such as surfactants, viscosity modifiers, residues from polymer synthesis, salts which are concentrated in the coating pores surrounding the microorganisms during film formation, may also be evaluated individually. Characteristic polymer synthesis residues present in a low T_g acrylic/vinyl acetate copolymer latex (for example Rovace™ SF091 ~300 nm average particle diameter, $T_g \approx 10°C$, Rohm & Haas) have been evaluated by our group to determine toxicity to a panel of bacteria, spores and yeast. These include vinyl acetate (1400 ppm), acetaldehyde (150 ppm), methyl propionate (130 ppm), butyl ether (160 ppm) and butyl propionate (65 ppm). Most Gram negative bacteria are not killed by these residue levels. Toxicity to vinyl acetate can be eliminated by using an all acrylate formulation. Some polymer synthesis residues however are toxic to Gram positive bacteria and yeast. Where toxicity is observed, dialysis (*38*), or ultra filtration can reduce toxicity, however, subsequent re-formulation may be necessary to maintain coating properties.

Commercial latex preparations contain anti-microbial biocides. For example, Rovace SF091 contains 15 ppm Kathon LX™, a mixture of isothiazolones (ACS Registry Nos. 26172-55-4, 2682-20-4), which loose potency over time and can be chemically inactivated prior to mixing with microorganisms. Chemical inactivation immediately prior to use is preferred because latex preparations without biocides can become contaminated with bacteria or fungi reducing their shelf-life. Most Gram negative bacteria added at the high concentrations used in biocatalytic coating formulations (~10^{11} colony forming units/ml emulsion) are not very sensitive to the presence of biocides added at a level of 15 ppm or 30 ppm. Gram positive bacteria and fungi may be more sensitive (Table I). Dialysis to remove toxic components may not remove all biocides because they may be

Table I. Acrylic Vinyl Acetate Copolymer Latex Emulsion Contact Toxicity to *Bacillus*, *Bacillus* spores, and Yeast

| Organism | % Loss (-) or Gain in Viability (cfu) | | |
	SF091 + Kathon	SF091, no Kathon	dSF091
B. subtilis 33608 (6)	-48±15	-53±10	-3±16
B. subtilis 33608 spores (3)	-11±5	-11±7	-10±10
B. subtilis spore outgrowth (3)	-20±10	-17±3	6±20
B. licheniformis 10716 (6)	-60±29	-60±20	-19±32
S. cerevisiae 204508 (5)	-30±14	-27±13	-20±14

SF091 latex: 15 ppm Kathon™, no Kathon™. Dialyzed SF091 latex: dSF091. Incubation: 30 min contact. (n), number of replicates.

weakly bound by polymer particle surface groups. In general, the density of microorganisms used in biocatalytic coating preparations is very high (50% v/v) and the number of cells killed by low levels of biocides can be a small fraction of the number originally added to the coating formulation.

In order to evaluate the effect of concentration of non-volatile polymer synthesis residues and biocides on the entrapped microorganisms, a film formation toxicity assay (FTA) was developed (*1, 57*). This assay measures cell viability during film (coating) formation as a function of emulsion composition, temperature, relative humidity, and the effect of substances concentrated in the coating during drying surrounding the microbes. In this method, a mixture of latex plus wet cell paste is rapidly coated into a 1 cm wide channel formed by a mask on a perforated polyester substrate. Coated 1 cm by 1 cm squares of polyester are periodically removed and the cells and non-coalesced latex immediately re-dispersed in buffer so that the viability of the re-dispersed microorganisms can be determined by viable plate counting (Figure 2). At the point of film formation, the coating cannot be re-dispersed and the cells are entrapped. The microorganisms detected in the buffer after film formation are released from the surface or the edges of the coating. Ideal coating formation is when all of the cells are instantaneously entrapped onto the substrate without loss of viability. However, entrapment by film formation is not ideal and some viability is lost due to contact toxicity prior to coalescence. For some microorganisms, significant viability is lost as a result of film formation if the coating has low porosity, is dried under low relative humidity, or when the microbes are osmotically stressed or starved for oxygen. Monitoring the change in cell viability as a coating is formed using the FTA assay can be combined with other measurements of microbial activity such as enzyme activity or the ability to carryout gene expression and protein synthesis. Transcriptional fusions of stress-inducible promoters with reporter genes (see below) such as *lux* (resulting in luminescence) or *gfp* (resulting in fluorescence) can give additional information about the effect of film formation on microbial physiology. Altered gene regulation can occur during film formation as a result of the change in water activity, partial cell desiccation and the changing chemical environment in the pore space.

Laboratory Patch and Strip Coating Techniques

A variety of Mayer rod draw down coating methods are used with pressure sensitive masks to make biocatalytic coatings on polyester or stainless steel in sheets, strips, or patches with edges sealed by a sealant nanoporous topcoat (*40, 43-45, 52, 53, 56, 57*) (Figures 3, 4). Filaments also can be coated (*58, 59*). Bilayer patch sizes for laboratory reactivity studies are 12 to 12.7 mm or 35 to 50 mm in diameter with coating thickness ranging from <10 μm to ~75μm (*52*). Re-hydrated patch volumes are <5 μl to ~125 μl. Strip sizes coated on polyester can vary from 1 cm^2 to 12.5 cm^2 (*56*). Essentially any thickness and size patch

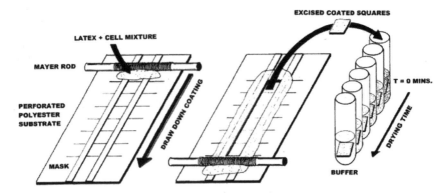

Figure 2. Latex film formation toxicity assay (FTA). Periodic removal of coated squares during film formation and drying in order to determine latex-entrapped cell viability. (Reprinted with permission from reference 1. Copyright 2007, American Chemical Society and American Institute of Chemical Engineers)

or strip can be made to entrap viable microorganisms as long as the drying temperature and relative humidity are controlled. Relative humidity during coat drying should be >50% for optimal retention of viability. At 50% volume fraction of wet cell paste, 12.7 mm and 35 mm diameter patches contain >1 x 10^8 and 1 x 10^{10} viable bacteria respectively, but this varies with coating thickness, cell size and shape. Without a sealant top coating of porous polymer, a significant fraction of some strains of microbes can be released from the patches by shaking in buffer depending upon the interaction between the cell and polymer particle surface. Latex blend coatings easily redispersible with mild sonication have recently been reported (*56*) in order to accurately determine the total number of viable cells in strips. When coated onto metals, latex coatings can be delaminated following rehydration and used as stand alone films or membranes (*44*). Because of the very high microbial density, the sealant top coat, and use in nutrient-limited non-growth media, coatings can be made under non-aseptic conditions (Figure 3).

Determination of Coating Nanoporosity and Pore Stability

Biocatalytic coating reactivity is directly related to coating thinness, microstructure and porosity which has been estimated in delaminated coatings by measurement of tracer diffusivity (*43, 44, 47, 48, 52, 53*). For estimation of coating porosity, latex coatings are cast on stainless steel and dried for 1 hour at 22-26°C, 50% relative humidity and delaminated in 5°C distilled water for 30 minutes. Coating on substrates such as photographic paper, lapping film (0.3 μm

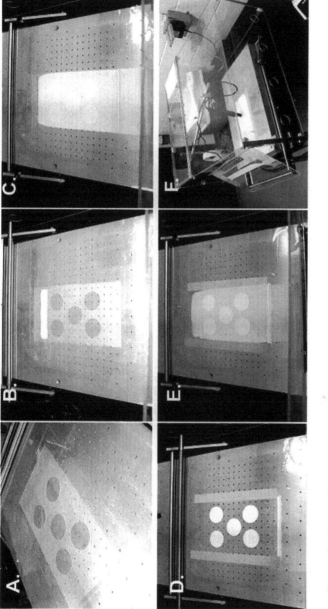

Figure 3. 35 mm diameter latex patch coating (52). A. Vinyl mask applied to polyester substrate on vacuum table. B. Cell coat emulsion applied to mask. C. Cell coat mixture coated over mask by Mayer rod. D. Removal of pressure sensitive mask. Application of spacer strips (opaque polymer used to show patches). E. Application of top coat emulsion by Mayer rod coating. F. Top coat drying in humidified chamber.

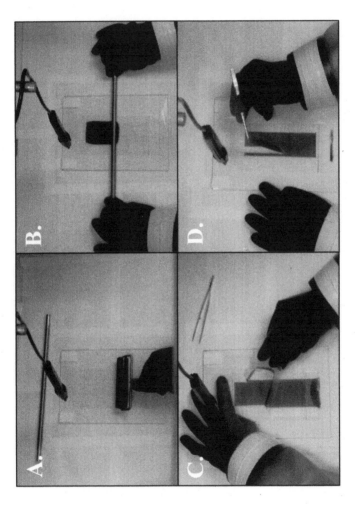

Figure 4. Strip coating of photosynthetic bacteria in an anaerobic cabinet with humidity control. A. Applying vinyl mask with a 4 cm x 12.5 cm opening to pre-cut polyester sheet substrate on glass plate. B. Mayer rod coating of microbial latex emulsion over mask. C. Removing pressure sensitive vinyl mask following coat drying. D. Removing a 1 cm x 12.5 cm coated strip to be inserted into a Balch anaerobe tube.

aluminum oxide), or coatings dried at low relative humidity (<40%) and delaminated from stainless steel had reduced permeability not linearly related to thickness suggesting that substrate and drying rate can lead to formation of anisotropic microstructure (*44, 57*). Coating thickness is measured with a digital micrometer. Stand-alone films 35 mm in diameter are punched from the delaminated coatings generated on stainless steel shims, mounted in a diffusion cell and the effective diffusivity of a tracer molecule with strong UV absorption (such as nitrate or riboflavin) injected into the donor cell and its accumulation measured in the receiver cell at 30°C with a flow through detector. The effective diffusion coefficient (D_{eff}) is calculated by the Nightingale's equation based on a pseudo-steady state approximation (*60*). The diffusivity of latex films is expressed as the ratio of D_{eff} and the diffusivity of the tracer in water (D). The diffusivity of latex coatings less than ~8 μm and coatings that could not be delaminated from stainless steel without tearing cannot be determined using this stand alone film method.

Several methods can be used to quench coalescence and generate coating porosity during film formation without killing the entrapped microorganisms such as decreasing capillary pressure by adding glycerol, coating at low temperature followed by a short high temperature "welding" of the polymer particles, or arresting polymer particle coalescence by the addition of porogens such as sucrose, sorbitol or trehalose. (*44*). Figure 5 shows the pore space generated by the addition of 0.37 g sucrose/g polymer solids to an acrylic/vinyl acetate copolymer latex (Rovace SF091, Rohm & Haas). Incorporation of this level of sucrose into the coating increases the diffusivity from $D_{eff}/D = 0.002$ to $D_{eff}/D = 0.072$ (Figure 6). At sucrose levels of 0.7 g/g latex, film formation with this emulsion is interrupted and the polymer particles redisperse (*44, 49, 52, 53*). Polymer particle surface groups such as hydroxyl ethyl cellulose may also affect the degree of coalescence and hence the porosity (*43, 52, 56*).

Cell containing coatings (0.3 to 0.4 v/v) with porogens (sucrose and glycerol) generally have D_{eff}/D values of ~0.07 to 0.11 (*44, 57*). The top sealant latex coat is a nanoporous sterile barrier that can have higher porosity but must be as thin as possible (without coating flaws) to reduce diffusion resistance and retain the cells in the coating (*53, 57*). The microbial cells themselves can significantly decrease porosity depending on their size, shape and volume fraction. *E. coli* mass fractions as high as 150 g cell dry weight/l of coating volume have been reported which reduced cell coat diffusivity to D_{eff}/D <0.05 (*44*).

Rehydration of acrylate/vinyl acetate copolymer coatings, which incorporate sucrose as a porogen to arrest polymer particle coalescence, results in dissolution of the entrapped vitrified carbohydrate from the pore space and continued wet coalescence may be significant. The temperature dependence of wet coalescence of low T_g SF091 coatings has been determined (*52*) described by an Arhenius relationship with an activation energy of 108kJ/mol (*50*) (Figure 7).

Figure 5. Cryo-SEM image of the top of an acrylate vinyl acetate copolymer coating (T_g ~10°C) with ~0.4 g sucrose/g polymer showing pore pace between partially coalesced polymer particles. (Reproduced with permission from reference 44. Copyright 2001 American Chemical Society).

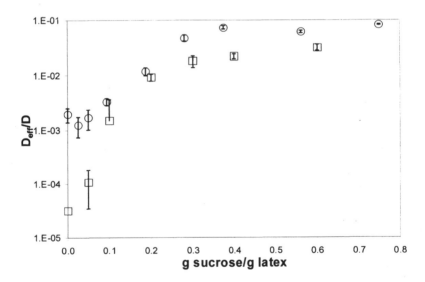

Figure 6. Relative nitrate diffusivity, D_{eff}/D, as a function of sucrose concentration in rehydrated Rovace SF091 coatings cast onto stainless steel and delaminated. Data from (52), O; data from (43), □. Error bars ±1 SD.

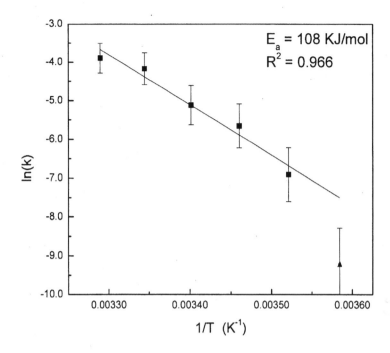

Figure 7. Loss of D_{eff}/D as a function of temperature during hydrated latex coating storage over a period of nine months (■). Equation: $k = Ae^{-Ea/RT}$, $A = 3.5 \times 10^{15} h^{-1}$, where k is the observed exponential decay rate of film permeability. Data from storage at $5°$ C was not used (▲); this temperature is below the T_g of $\sim 10°$ C (52).

Arresting wet coalescence of rehydrated latex coatings can be accomplished by using a bimodal polymer blends, core shell or high aspect ratio polymer (HARPS) blends. The diffusivity of polymer blends and core shell coatings is a function of blend ratio and shell thickness respectively (49, 50). The addition of sucrose to a bimodal blend with 800 nm polystyrene particles does not alter coating permeability indicating that formation of carbohydrate glasses in the pore space does not contribute to nanoporosity (50, 52). Using an 800 nm polystyrene particle (T_g = 94°C) mixed with 20% (v/v) of an 158 nm acrylate/styrene particle (T_g = -5°C), a bimodal blend coating was developed that can be cast at 30°C and has stable porosity at 80°C in sea water for entrapment of the marine hyperthermophile *Thermotoga maritima* (Figure 8). Starch hydrolysis by the toga-associated amylase of *T. maritima* at 80°C was stabilized for over 200 hours by concentration and entrapment in this adhesive bimodal blend coating compared to suspended cells which lost activity in less than 24 hours (43, 50).

Imaging Coating Microstructure and Nanoporosity

Cryogenic scanning and field emission scanning electron microscopy (cryo-SEM, cryo-FESEM) images of coating microstructure pioneered by Davis, Scriven, and coworkers confirm arrested polymer particle coalescence and reveal extensive nanoporosity surrounding the entrapped microorganisms (*43, 47, 48, 61, 62*). The large dark pores in Figure 9 are spaces occupied by the bacteria before they were released from the coating and in general the size of these spaces does not differ significantly from the size of the microorganisms measured in suspension (*44*). The nanopores formed by arresting coalescence of these polymer particles are <20 nm in diameter (Figure 9).

Lyngberg (*44*) also correlated the observed pore structure of the top of a coating using image analysis with diffusivity measurements. Coatings cast on silicone wafers and plunged into liquid nitrogen can be broken on a cryo-stage to produce an image of the entire cross section of the cell coat/topcoat interface or the depth of the entire coating (Figure 10) (*52, 53, 57*).

High magnification cryo-FESEM images of the top surface and fracture surfaces show partially coalesced polymer particles and ~50 nm fissures (Figure 11A, B). Cryo-fracture surfaces show characteristic star-shaped polymer particle ductile pull-outs observed by Ma (*49*) and Ge (*63*) (Figure 11B, C) as well as the structure of the pore space surrounding the entrapped microorganisms in the fracture plane (Figure 11C).

The cross section of multi-layered coatings also can be visualized using low magnification fluorescence microscopy of coating slices with pigmented layers (*43*). While this technique does not visualize individual polymer particles, embedded microorganisms or pore nanostructure, it is useful to detect coating quality, thickness, and flaws, especially in the sealant topcoat at the patch edge, that can result in release of cells into the medium (Figure 12) (*45*). Because laboratory scale coatings are small, edge effects on multilayer coatings can be significant. Alternative coating methods using a second mask have been suggested to minimize these coating edge effects on small bilayer patches (*57*).

Model Systems for Engineering Coating Reactivity

Gluconobacter oxydans Coatings for Oxidation of D-Sorbitol to L-Sorbose

A challenging model system for casting coatings of "pre-activated" microbial cells is engineering a bilayer latex coating to maintain the viability and enzyme activity of strictly aerobic microorganisms during film formation, coat drying and storage. Acetic acid bacteria such as *Gluconobacter* are used in industry for quantitative partial oxidation of polyols to ketones, ethanol to acetic acid, sugars to acids (*64*) and are strict aerobes sensitive to oxygen starvation (*65*). For this reason, methods have recently been developed to make model

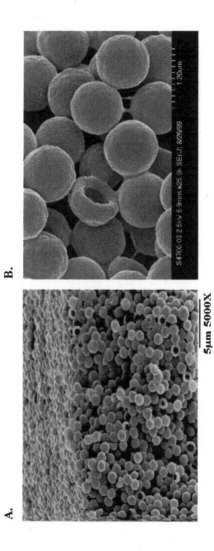

Figure 8. Cryo-SEM images of a bimodal blend latex coating of 0.8 (v/v) 800 nm polystyrene (Ropaque 1055) and 0.2 (v/v) 158 nm acrylate/styrene (JP1232) dried at 30°C and stable at 80°C in sea water. A. Top view and fracture surface. B. View of fracture edge. (Reproduced with permission from reference 50. Copyright 2005 Springer-Verlag).

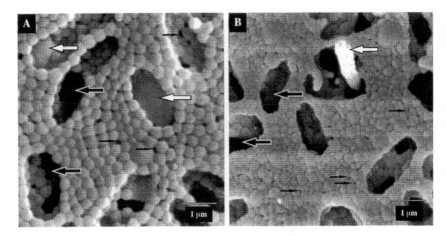

Figure 9. Cryo-SEM images of coating voids (large black arrows) created by entrapment of E. coli and pore spaces between polymer particles (small arrows). White arrows: E. coli. A. Coating generated with monodispersed acrylate/vinyl acetate copolymer latex with sucrose and glycerol. B. Coating generated with polydispersed acrylate/vinyl acetate copolymer plus glycerol (Reproduced with permission from reference 48. Copyright 1999 Elsevier).

latex coatings of *Gluconobacter oxydans*, a Gram negative aerobic bacterium which is used for the oxidation of D-sorbitol to L-sorbose, a biological step in the process of L-ascorbic acid (vitamin C) production. This oxidation is carried out by a membrane-associated sorbitol dehydrogenase (*64*).

The kinetics and effectiveness factor (η, the ratio of the observed coating-entrapped reaction rate and the reaction rate of suspended cells without diffusion limitations) of the quantitative oxidation of D-sorbitol to L-sorbose by *G. oxydans* was determined in a non-growth nitrogen-limited medium composed of sorbitol, pyruvate and phosphate buffer to eliminate growth of released cells (*52, 53, 59*). This chiral oxidation was investigated using individual 6 µl volume 12.7 mm diameter acrylate/vinyl acetate bi-layer nanoporous latex patches as a function of cell coat and top coat thickness (Figures 10, 11 above). The activity of individual patch coatings was studied in agitated Petri dishes or oxygenated 5ml micro bioreactors calibrated for oxygen transfer (*52, 53, 57, 66*). With top coatings of greater than 12 µm thicknesses, <0.05 % of the cells were released from patches during the course of the experiments. The oxidation rate stopped when the patches were removed indicating that the small percentage of released nitrogen-starved cells did not contribute to the observed reaction rate (Figure 13).

This system allowed accurate determination of the effectiveness factor from only coating-entrapped *G. oxydans* as a function of coating thickness and the availability of oxygen. Even though the coatings were optimized for porosity, the viability of this strict aerobe in latex patches was reduced to 51 % even for

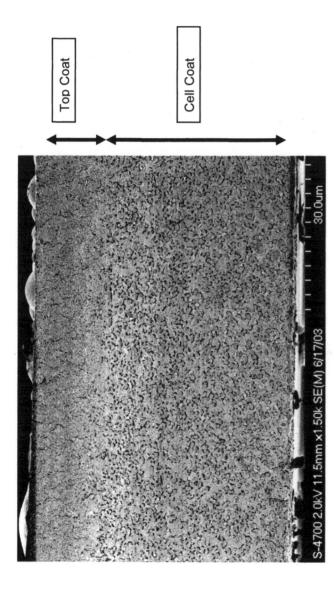

Top Coat

Cell Coat

Figure 10. Freeze-fracture cryo-SEM image of a Gluconobacter oxydans bilayer coating. Cell coat thickness 35 μm; topcoat thickness 15 μm (Reproduced with permission from reference 53. Copyright 2006 John Wiley & Sons).

A.

B.

C.

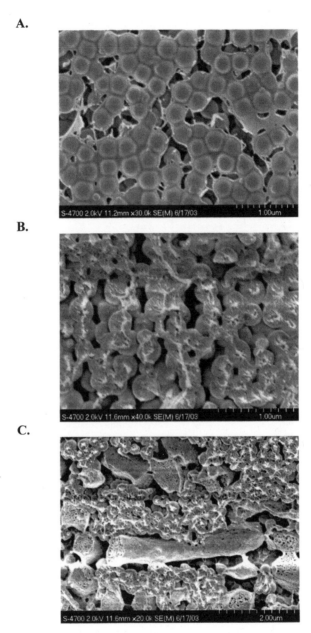

Figure 11. Cryo-fractured SEM image of a Gluconobacter oxydans coating (52, 57). A. Top coat surface showing fissures. Cryo-fracture of top coat showing fissures and polymer particle pull-outs. C. Cell coat showing pore space surrounding latex-embedded and cryo-fractured G. oxydans.

A.

B.

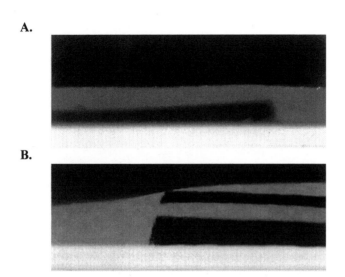

Figure 12. A. Fluorescence microscopy side-view of a slice of the edge of a bi-layer latex patch showing the slightly concave cell coat and reduced top coat thickness at the edge. B. Side-view of a portion of a 4-layer patch (3 patch coatings, 1 top coat) showing a coating flaw at the edge where the sealant top coat does not completely cover the top layer. Substrate (bright layer) 125 μm polyester. (Reproduced from reference 43.)

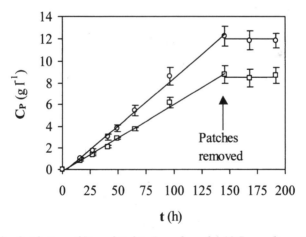

Figure 13. Oxidation of D-sorbitol to L-sorbose by 12.7 mm diameter bilayer patches of latex entrapped G. oxydans. ○: cell coat thickness 21 ± 2 μm, top coat thickness 13 ± 1 μm; □: cell coat thickness 12 ± 2 μm, top coat thickness 17 ± 3 μm. C_p, L-sorbose concentration. N = 6. (Reproduced with permission from reference 53. Copyright 2006, John Wiley & Sons).

cell coat thicknesses of <10 μm, indicating the sensitivity of this microbes to oxygen starvation during film formation and drying. The addition of the sealant polymer topcoat layer further reduced viability from 51% down to 30 % and 11 % as the nanoporous top coat thickness was increased from 9 μm to 20 μm respectively. In spite of this low cell viability, very high reaction velocities (intensity) as a function of coating volume were observed from *G. oxydans* patch coatings, 84 and 102 g/l coating volume · h (*53*).

A bilayer diffusion reaction model was developed by Charaniya and Fidaleo (*53, 57*) to predict η as a function of cell coat thickness and sealant top coat thickness in order to optimize cell concentration, cell coat thickness, top coat thickness, and to predict the oxygen concentration profile and oxygen consumption rate in the cell coat layer as a function of bulk liquid oxygen concentration. Simulations using the diffusion reaction model and the experimental data (Shown in Figure 13 above) indicate that extremely high coating reactivity for oxidation of D-sorbitol to L-sorbose can be achieved, especially using coatings of <10 μm thickness. L-sorbose volumetric production rates of >100 g/l coating volume · h can be achieved (Figure 14) if both D-sorbitol and oxygen can be continuously supplied at a sufficient rate to not limit the reaction velocity. The observed effectiveness factors were higher than the model-predicted values of η because of the non-linear dependence of viable *G. oxydans* density in the cell coat as a function of oxygen availability resulting from the diffusion resistance of the top coat. In spite of the much lower porosity of bilayer latex patches, the observed latex coating effectiveness factor (η = 0.22) was >20-fold higher than previously reported by Müh et al. for this same immobilized *Gluconobacter* oxidation (*67*).

Further work is needed to develop adhesive latex emulsions that generate more porous coatings and coating methods that do not reduce the viability of strict aerobes such as *G. oxydans* during film formation and coat drying, particularly for multi-layer coatings. However this study confirms that very thin microbial biocatalytic coatings can be engineered for very high volumetric reactivity (intensity), which could achieve much higher rates than biological oxidation systems currently used in industry.

The ability of nanoporous latex coatings to stabilize D-sorbitol oxidation activity was determined in bilayer 12.7mm diameter *G. oxydans* latex patches made with glycerol and 0.4 g sucrose/g polymer which each contained 1.1 to 1.3 x 10^8 viable cells (*52, 57*). Oxygen was not limiting the reaction rate. The half-life of latex-entrapped cells for the decay in the rate of oxidation of D-sorbitol to L-sorbose in non-growth media was ~430 hours compared to only ~28 hours for suspended cells (Figure 15). (The low reaction rate compared to Figure 13 is due to a 56 μm sealant topcoat which resulted in release of <0.05% of the viable cells. The oxidation rate could be significantly increased with a thinner top coat, *57*) The reactivity of the latex-entrapped *G. oxydans* decreased linearly with time compared to an exponential decay in reactivity for suspended cells (Figure 15A) which is due to the loss of some of the suspended cells with each subsequent centrifugation and re-suspension in fresh medium.

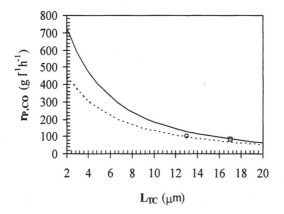

*Figure 14. Simulated and experimental L-sorbose volumetric production rate
($r_{P,CO}$) in bilayer latex coatings as a function of top coat thickness (L_{TC}) at
different cell coat thickness (L_{CC}).* ------ (L_{CC} = 12 μm), - - - - - (L_{CC} = 21 μm)
*calculated by a diffusion reaction model. Symbols: ◯, ▢, oxidation rate data
from Figure 12 above. (Reproduced with permission from reference 53.
Copyright 2006, John Wiley & Sons).*

Although the accumulation of L-sorbose was faster for suspended cells
during the initial 150 hours due to the absence of a diffusion barrier and higher
viability of suspended nongrowing cells (Figure 15A), D-sorbitol oxidation and
L-sorbose accumulation by latex coating entrapped *G. oxydans* continued for
over 600 hours (*52*). *G. oxydans* patches could be stored for at least seven
weeks at 4°C without loss of reactivity.

Gene Expression Following Rehydration: Reactive Coatings as Biosensors or to Detect Microbial Stress during Film Formation

*What promoters are useful for regulating gene expression in nongrowing latex-
entrapped E. coli?*

In contrast to strict aerobes, both anaerobic and facultative anaerobic
bacteria such as *E. coli* retain high cell viability and reactivity in latex coatings
(*39, 41, 42, 50*). Investigation of nongrowth associated gene expression, protein
synthesis and *in vivo* protein activity has been carried out in latex-entrapped *E.
coli* because of the ease of constructing plasmids encoding promoters fused to
indicator genes such as *lacZ*, *lux* or *gfp* which respectively encode β-
galactosidase, luciferase or green fluorescent protein (GFP). Using gene
fusions, the kinetics of nongrowth associated inducible gene expression can be
monitored non-destructively in coating patches using color changes (such as β-

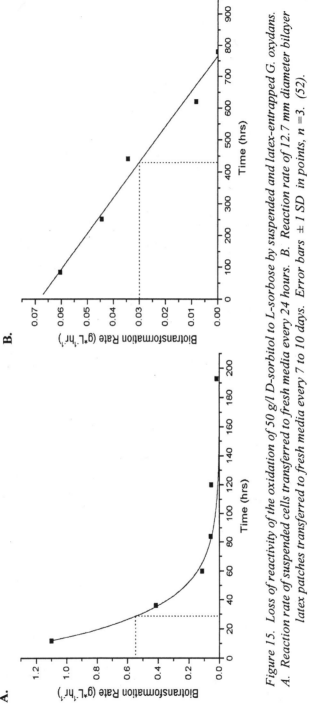

Figure 15. Loss of reactivity of the oxidation of 50 g/l D-sorbitol to L-sorbose by suspended and latex-entrapped G. oxydans. A. Reaction rate of suspended cells transferred to fresh media every 24 hours. B. Reaction rate of 12.7 mm diameter bilayer latex patches transferred to fresh media every 7 to 10 days. Error bars ± 1 SD in points, n = 3. (52).

galactosidase cleavage of a non-colored substrate to produce a colored compound), luminescence or fluorescence. The p*lac* lactose or IPTG-inducible promoter was initially found to up-regulate β-galactosidase expression in *E. coli* coatings incubated in a nitrogen-starvation buffer (*39*). Subsequently it was found that latex-entrapped *E. coli* contains both σ^{70} and σ^{S} RNA polymerase forms (Figure 16, *68*) suggesting that promoters regulating gene expression either during vegetative growth or during stationary phase might be useful in latex-entrapped non-growing cells.

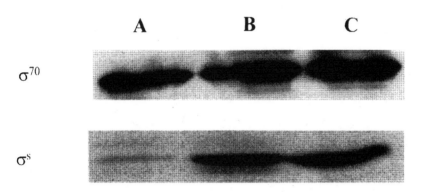

A   **B**   **C**

σ^{70}

σ^{S}

Figure 16. RNAP σ forms present in latex coatings of E. coli K211 (Western blot using anti σ^{70}, σ^{S}). A. Mid-exponential growth. B. Stationary phase. C. Latex-entrapped stationary phase E. coli (68).

E. coli promoter-fusions useful for monitoring gene expression in coatings

Of the promoters investigated for nongrowth associated gene expression in latex coatings, the ionic mercury-inducible *mer* operon promoter was especially active in latex-patches (100,000-fold up regulation). This lead to subsequent development of a single-use latex patch biosensor for bioavailable ionic mercury (*46*) using *E. coli* HB101 containing the plasmid pRB28 encoding a mercury-inducible *merRo/pT'-lux* (*mer-lux*) fusion developed by Selifonova (*69*).

Recently, a *mer-gfp* green fluorescent protein reporter has been constructed (*70*) to nondestructively study the kinetics of spatial gene expression in a 35 μm thick nanoporous bilayer coatings using laser scanning confocal microscopy. To orient the location of the GFP-expressing *E. coli* cells within the coating, fluorescent Estapor@Y microspheres (555/570 nm excitation/emission) were added to the topcoat, and Flash Red fluorescent microspheres (660/690 nm excitation/emission) were added to the cell coat. Because of the different excitation and emission properties of these microspheres relative to GFP (488/509 nm excitation/emission), all three fluorescent molecules could be detected and quantified simultaneously. The analysis indicated that the *E. coli* closest to the topcoat-cell coat interface showed a higher level of GFP induction relative to cells positioned closer to the polyester substrate (Figure 17). Higher concentrations of mercury resulted in increased GFP fluorescence in the bottom layers of the coating suggesting that inducer (Hg^{+2}) diffusion limits patch reactivity. Z-plane images of fluorescence intensity of X-Y optical "slices" can be used to quantify spatial expression as a function of depth into the coatings. Top coat thickness did not alter spatial GFP expression (Figure 18).

Laser scanning fluorescence confocal microscopy is a valuable tool for quantifying spatial gene expression within microbial latex coatings, and the use of fluorescent microspheres allows for precise optical localization of gene expression (microbial reactivity) within each coat layer. This system can be used to optimize latex coating reactivity as smart biosensors and biocatalysts with regard to cell coat and top coat thickness, the effect of nanoporosity and pore wall chemistry on the diffusion of oxygen, nutrients and inducers through the coatings.

In addition to biological fluorescence, microbial luminescence is a useful tool to detect microbial viability and reactivity in latex coatings. Luminescence from latex patches can be detected in a luminometer or by a scintillation counter. *E. coli* stress-inducible promoters fused to *lux*CDABE lux operon have been evaluated along with the *mer* ionic mercury inducible promoter under non-growth nitrogen-limited conditions in latex coatings with different chemical "stresses" including temperature (25°C and 37°C) (*71, 72*). Some of these promoters are recognized by RNAPσ[70] and some recognized by RNAPσ[s] (*71*). The different kinetics of luminescence induction, intensity and duration, as well as response to temperature by different promoters fused to the lux operon indicates a potentially broad range of selective reactivity of microbial smart coatings (Figure 19) (*72*). Coatings can be formulated containing either a single microbe or a mixture of microbes engineered for differential sensitivity to a wide variety of compounds in the environment. The only difference would be the promoter sequences in the plasmid DNA of the entrapped strains. These results indicate that microbial smart coating biosensors that respond to temperature, salt, toxic metals, oxidizers (H_2O_2), ethanol, DNA-damaging agents, oxidative conditions, and other changes in their chemical environment (osmotic shock, desiccation, the presence of toxins, etc.) can be constructed by selecting appropriate promoters. Patch, pattern coating, multilayer or mixtures of different microbes in a single layer coating methods may be used to combine differentially-responding microbes.

Latex ink-jet printed microstructures containing reactive E. coli

Using the ionic mercury-inducible luminescent *E. coli* strain, miniature arrays of Hg^{+2}-responsive latex microstructures were fabricated using ink-jet

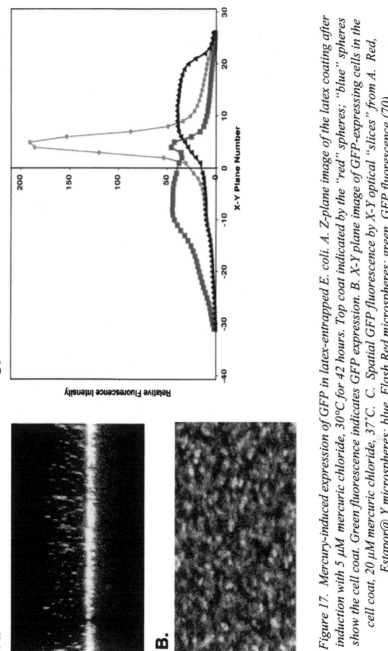

Figure 17. Mercury-induced expression of GFP in latex-entrapped E. coli. A. Z-plane image of the latex coating after induction with 5 μM mercuric chloride, 30°C for 42 hours. Top coat indicated by the "red" spheres; "blue" spheres show the cell coat. Green fluorescence indicates GFP expression. B. X-Y plane image of GFP-expressing cells in the cell coat, 20 μM mercuric chloride, 37°C. C. Spatial GFP fluorescence by X-Y optical "slices" from A. Red, Estapor@ Y microspheres; blue, Flash Red microspheres; green, GFP fluorescence (70).

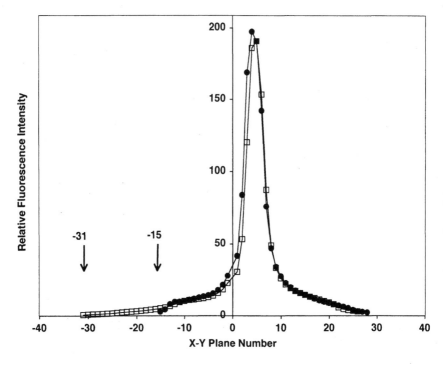

Figure 18. Comparison of spatial GFP fluorescence intensity as a function of coating depth and top coat thickness. Coating conditions as in Figure 16 above. Top coat thickness: □, 28 µm; •, 13.5 µm (70).

printing (Figure 20) (*13, 43*). Latex emulsions were developed for ink-jet deposition of viable *E. coli* through a variety of piezoelectric tip orifices (*13, 43*). The ink-jet deposited microstructure dots measured by optical microscopy and profilometry indicated formation of craters resulting from substantial surface tension driven flow prior to drying (Figure 20A, B). Surprisingly, latex emulsions of *E. coli* survived ink-jet deposition and rapid drying and the cells were responsive to ionic mercury (Figure 20C). Using this same approach, miniature array sensors of different microbes could be printed to detect the composition of contaminants in a complex fluid (liquid or gas) by detection of differential gene expression.

Engineering Photo-Reactive Coatings of *Rhodopseudomonas palustris* for Optimal Light Adsorption and Hydrogen Production

A valuable application of microbial smart coatings is to preserve and uniformly illuminate photosynthetic microbial, cyanobacteria or algal cells to

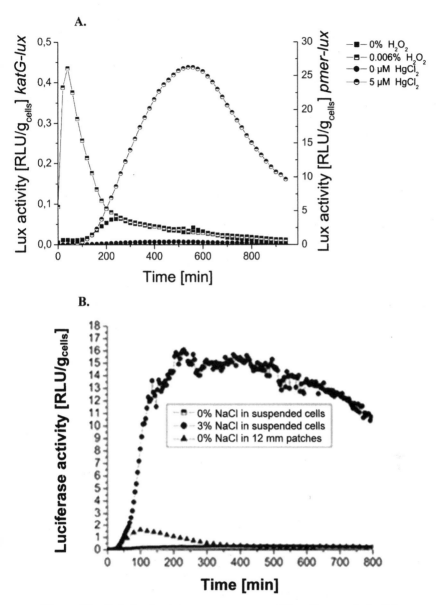

*Figure 19. Relative luminescence response of E. coli stress-inducible
promoters fused to lux in 12 mm diameter latex patches under non-growth
conditions (nitrogen-limited, n = 3). A. pkatG-lux at 37°C: ▲ 0% H₂O₂,
▼ 0.006% H₂O₂; pmerRo/pT'-lux at 25°C: ■ no Hg⁺², □ 5 μm Hg⁺².
B. pxza-lux inducible with NaCl at 37°C (72). (Reproduced with permission
from reference 1. Copyright 2007 American Chemical Society and American
Institute of Chemical Engineers)*

A.

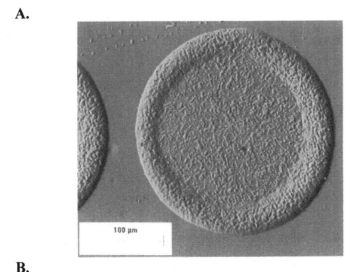

B.

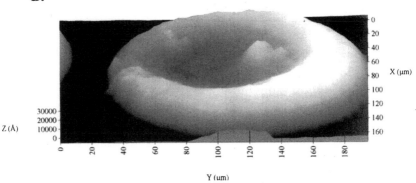

Figure 20. Ink-jet deposited E. coli pmerR-lux using an acrylate/vinyl acetate latex ink formulation (13). A. Dot printed using 50 μm diameter piezo tip, ~ 9 x 10⁴ E. coli per dot, ~170 μm dot diameter. B. Profilometer image of E. coli latex dot. Thickness: center ~2 μm, rim 6 -10 μm. C. Luminescence response of 10 x 10 dot arrays of printed E. coli induced with 100 nM Hg⁺²: ●, 250 nl total volume (2.5 nl per dot) printed with 50 μm tip, ■, 100 nl total volume (1 nl per dot) printed with 25 μm tip (13). (C. Reproduced with permission from reference 1. Copyright 2007 American Chemical Society and American Institute of Chemical Engineers)

C.

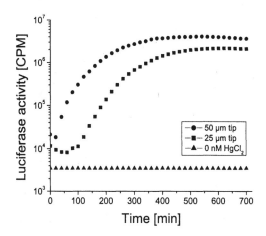

Figure 20. Continued.

generate photo-reactive paints and inks. This was first reported for *Synechococcus* by Martens & Hall (*38*). Nitrogenase activity under non-growth conditions of the metabolically versatile photosynthetic bacterium *Rhodopseudomonas palustris* CGA009 (*73*) is being evaluated in acrylate/vinyl acetate latex strip coatings (see Figure 4 above) in micro photobioreactors with an argon atmosphere (Figure 21) (*56*). Under these conditions *Rps. palustris* catalyzes the photo-decomposition of acetate (and other organic acids) with generation of hydrogen gas and carbon dioxide.

Single layer, ~65 μm thick 6.25 cm^2 coatings of *Rps. palustris* containing ~10^{10} cells produce more hydrogen per m^2 than the same quantity of settled cells. Hydrogen evolution is light dependent (Figure 22) and stable at 30°C. Latex-entrapped *Rps. palustris* strips can be stored frozen at -80°C for one year without loss of photo-reactivity, and coating microstructure (porosity, the presence of osmoprotectants, glycerol, sucrose) affects hydrogen production rate (Figure 23).

The genome sequence of *Rps. palustris* CGA009 encodes three functional nitrogenase with different reactivity (*73*) and a mutated uptake hydrogenase (*74*). This coating system along with genetic engineering of *Rps. palustris* can potentially can used to engineer optimal biological photon adsorption and hydrogen evolution per meter surface area using multiple layers containing different nitrogenase and photo-pigment mutant strains optimized spatially for light adsorption by wavelength and intensity (Flickinger *et al.*, PCT WO 2005/014805 A1, US 2005/0176131 A1). These photoreactive coatings may be useful for generation of hydrogen for fuel cells or as inexpensive biological photo-adsorbers for generation of hydrogen for hydrogenation reactions.

84

*Figure 21. 1 cm x 6.25 cm² nanoporous coating of latex-entrapped Rps.
palustris CGA009 in an argon atmosphere which produces H₂ (56).*

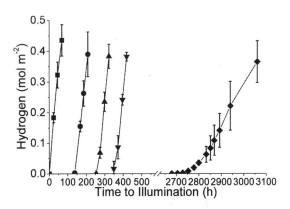

*Figure 22. Hydrogen production by latex coatings of Rps. palustris CGA009 as
a function of storage without light (foil covered tubes) at 30°C. Illumination 32
μmol photons/m² · s. Error bars ± 1 SD, n=3. (Reproduced with permission
from reference 56. Copyright 2007 American Chemical Society and American
Institute of Chemical Engineers)*

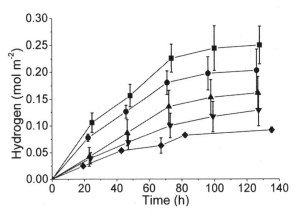

*Figure 23. Hydrogen production from Rps. palustris latex strips in an argon
atmosphere, 20 mM acetate. Coating formulations contain: ■, glycerol + sucrose;
●, sucrose only; ▲ glycerol only; ▼, no porogens; ♦, settled cells. Error bars ± 1
SD, n = 3. (Reproduced with permission from reference 56. Copyright 2007
American Chemical Society and American Institute of Chemical Engineers)*

Coating *Geobacter sulfurreducens* on Electrodes for Electrochemistry and Microbial Fuel Cell Studies

Microbial coating technology may also be a useful tool to study microbe-electrode interactions and to harness bacteria as current-generating or current-directed biocatalysts. It has recently been reported that anaerobic bacteria capable of utilizing metals as electron acceptors can also utilize electrodes as electron donors or acceptors (*75-77*). The most common approach to study this interaction is a carbon electrode suspended in a reactor inoculated with bacteria. Characteristics of this 'fuel cell' are monitored over days to weeks, following attachment and growth. Deposition of adhesive nano-porous thin coatings of bacteria could avoid the need to pre-colonize electrodes, and create uniform microbial electrodes for controlled studies. In addition, entrapping cells would minimize release of bacteria under bioprocessing conditions, where biomass in the effluent is undesirable.

Geobacter sulfurreducens produces fragile extracellular pili structures, thought to be required for attachment to substrates and electron transfer to electrodes (*78*). Adhesive latex and composite coating methods are currently being investigated for coating *Geobacter* on conductive materials, however many latexes are toxic to *Geobacter* and this microbe is sensitive to drying rate (*79*). Preliminary studies of cultures of *G. sulfurreducens* immobilized on graphite paper using 4% pectin (cross linked with $BaCl_2$) immediately demonstrated electron transfer to the electrode, as monitored via cyclic voltammetry and electrochemical impedance spectroscopy (Figure 24) (*79*). Subsequent incubation of *Geobacter*-coated graphite electrodes demonstrated sustained electrical output, rising to 1.75 A/m^2 within 24 hours. Electrochemical features of these coatings were identical to electrodes colonized in a microbial fuel cell for weeks (*79*). Further investigation of microbial electron transfer phenomena and development of latex-based polymer coatings that immobilize bacteria while allowing them access to insoluble minerals or conductive electrodes could lead to development of current-generating microbial coatings for directed biocatalysis or conversion of waste organic materials into electricity.

Microbial Smart Coatings:
Challenges and Future Applications

A variety of latex microbial smart coatings are being investigated as model systems in order to determine the fundamental properties for engineering highly reactive adhesive paints and inks containing living cells. These include coatings for chiral oxidations, reductions, photo-reactive coatings, coatings that produce hydrogen, coatings that produce electric current, coatings of extremophile microorganisms stable in harsh temperature and chemical environments, and very thin coatings/microstructures generated by ink-jet deposition. This requires a

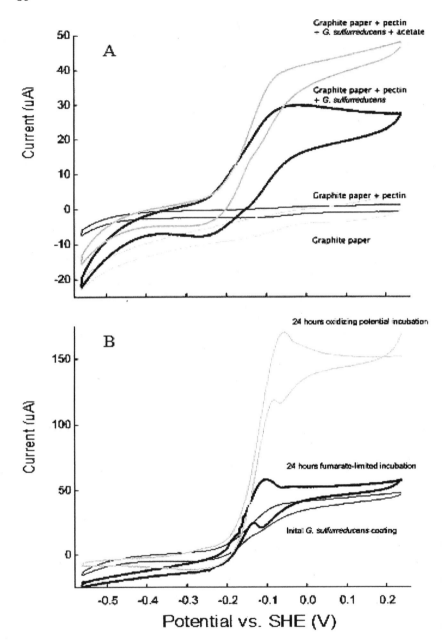

Figure 24. Cyclic voltammetry (10 mV/s) of graphite paper electrodes coated with Geobacter sulfurreducens and cross linked pectin. A: Initial current production, B: effect of 24 h incubation with the electrode poised at +0.25 V, or in fumarate-limited medium (79).

cross-disciplinary team of investigators in polymer chemistry, coating technology, chemical and materials engineering, and microbial physiology. Of primary concern are: coating nanoporosity, the cellular response to desiccation during film formation, regulation of microbial gene expression in cells that are not growing, reactivity (intensity) and kinetics, photo-adsorption and light scattering, and cellular preservation/stabilization by vitrification of carbohydrate glasses. Still to be explored are smart coatings which stabilize microorganisms with non-growth associated biosynthetic capabilities to generate nanoporous paints and inks which secrete useful products. This includes coatings that secrete antibiotics, regulatory metabolites, peptides, enzymes and biological corrosion-inhibiting coatings.

Understanding the Chemical Environment, Distribution and Structure of Water and Other Molecules Concentrated in the Dried Pore Space

Critical to future development of microbial smart coating technology is the molecular understanding of preservation of microbial viability at ambient temperature when microbes are entrapped in adhesive nanoporous coatings. The science of biopreservation was started with the discovery of higher than normal levels of certain molecules in the cytoplasm of desiccation and freeze-tolerant species. When these organisms are exposed to gradual changes in temperature, salinity or humidity, osmolytes are accumulated (ions, proteins, carbohydrates, amino acids or urea) which helped them survive the harsh conditions (80). In desiccation tolerance, which is relevant to latex film formation, the role of intracellular osmolytes is not known, but they likely modify the activity and manipulate the state of intra- and extracellular water. It is hypothesized that they either replace structural water in the immediate vicinity of proteins and membranes (*81*) enabling these structures to maintain their native configuration or are preferentially excluded from the surface of proteins and membranes to help preserve structural water (*82*).

Much literature exists on preservation of biological molecules and viable cells by glycerol, by vitrification of carbohydrates to produce extracellular glasses, the effect of salts concentrated during drying, and processing parameters (drying temperature, relative humidity, drying rate, drying time) (*81*). It is also believed that this same biomolecular structural transition from a low viscosity liquid to a glassy state occurs within living cells when dried, freeze-dried or frozen (*81*). In addition, microbes such as *E. coli* respond to desiccation and osmotic stress by synthesis of cytoplasmic osmolytes such as glycine betaine (*83-85*), and yeasts produce trehalose (*86*) so the physiological response of the latex-entrapped microbes themselves may be the most sensitive measure of how to optimize coating microstructure for retention of viability and reactivity.

For the particular case of confinement of biomolecules and water in coating nano-pores where polymer surface effects dominate, it is likely that water activity is altered. This is due to the dipolar nature of water, which causes it to change its residence time as a function of its immediate environment (*87*). For example, the

88

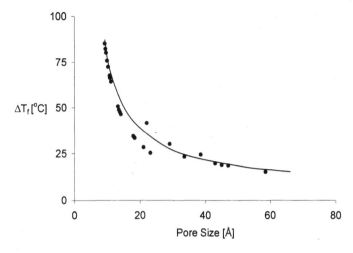

*Figure 25. Change in the freezing temperature of water as a result of
confinement in nanopores. Compiled from data from references 88 and 89.
(Reproduced with permission from reference 1. Copyright 2007 American
Chemical Society and American Institute of Chemical Engineers)*

freezing temperature of water when confined in nanopores decreases with pore
size (Figure 25) (*88, 89*). Confinement in coating nanopores has the same effect
as exposure to osmolytes, with respect to water activity and modification of
biochemical reaction rates. The phase change behavior (*87*), T_g and crystallization
rates of solutions are also affected by confinement (*90, 91*).

These changes in the mobility of water and solvents enables stabilization of
biological materials in nanoporous coatings resulting from increasing fluid
viscosity and entropic confinement during evaporation and simultaneous pore
formation by polymer particle coalescence. A microfluidic device has recently
been reported by Aksan as a tool for directly measuring the change in extracellular
viscosity and microbial membrane thermodynamic behavior (membrane phase
transition) as a function of water content during vitrification (*92*).

The Physiology of Non-Growing Microorganisms is Poorly Understood

In addition to the effects of water activity during film formation, the biology
of latex-embedded microorganisms that are not growing (nutrient-limited) or
growing very slowly following coating rehydration has yet to be investigated.
This includes the redox and energy balance in non-growing cells, DNA repair
mechanisms active in the absence of DNA synthesis, ribosome stability, RNA
polymerase and mRNA stability (which regulate gene expression), regulation of
protein turnover (intracellular degradation by proteases) and new protein
synthesis which effects *in vivo* protein stability. By understanding how to

manipulate the regulation of gene expression without growth, the reactivity of biocatalytic coatings could be increased by several orders of magnitude. In addition, understanding the regulation of *in vivo* proteolysis in nitrogen-limited non-growing microorganisms could lead to strategies to alter protease activity in latex-entrapped cells and the design of proteins with exceptionally long *in vivo* functional half-lives.

Coating and Ink-Jet Printing Methods are Needed for Generating Highly Structured or Perfusive Microbial Smart Coatings

Of parallel importance with increasing the specific reactivity and active half-life of the embedded microbial biocatalysts is the development of combinations of both nanoporous and nonporous latex with nano-structured inorganic matrices or microchannels to produce mechanically stable coatings with reduced toxicity, increased porosity, and light penetration (optical properties) (*93*). Development of mechanically strong latex perfusive coatings, highly structured coatings with microchannels, wash coatings for monolithic microchannels (Figure 26) and methods to generate stand-alone membranes that retain porosity for very long periods of time, even when exposed to extreme temperature and reaction conditions, is essential to further reduce diffusion resistance and increase smart coating reactivity.

An extension of this smart coating technology is miniaturization by fabrication of very small coatings and latex microstructures containing highly reactive living microorganisms by improving formulations of latex microbial inks for piezoelectric droplet deposition (*13*). Using printing technology, precise deposition of a high density of living microorganisms onto miniature bionano devices and the sensing elements of integrated circuit chips (ICs) is possible so that the metabolism of microorganisms can be incorporated into the design of bioelectronic devices.

Acknowledgements

The preparation of polymer formulations and collaboration of Dr. Matthew Gebhard, and Kathie Koziski, Rohm & Haas, Spring House, PA is gratefully acknowledged. Elizabeth "Za" Freeman assisted portions of the ink-jet deposition studies. Portions of this work were supported by the University of Minnesota BioTechnology Institute, the University of Minnesota NIGMS Biotechnology Training Grant, the National Science Foundation Award DBI-0454861, the NSF National Nanotechnology Infrastructure Network REU summer program, DSO/DARPA contract N66001-02-C-8046, the University of Minnesota Initiative for Renewable Energy and the Environment (IREE), Agricultural Experiment Station, and the National Research Council of Italy, Short-Term Mobility Program.

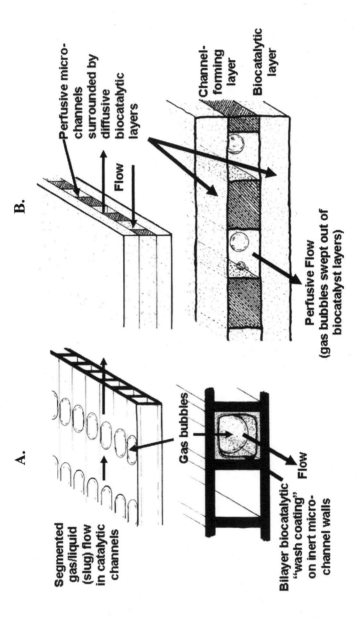

A.

Segmented
gas/liquid
(slug) flow
in catalytic
channels

Gas bubbles

Flow

Bilayer biocatalytic
"wash coating" on inert microchannel walls

B.

Perfusive micro-
channels
surrounded by
diffusive
biocatalytic
layers

Flow

Channel-
forming
layer

Biocatalytic
layer

Perfusive Flow
(gas bubbles swept out of
biocatalyst layers)

*Figure 26. Applications of latex smart coatings for microchannel and structured perfusive coatings for multi-phase biocatalysis.
A. Wash coating on the walls of a monolithic support for multiphase (gas/liquid/solid) biocatalysis with intermittent slug flow.
B. A perfusive structured coating with channels for rapid liberation of gas bubbles generated in the coating.*

References

1. Flickinger, M. C.; Schottel, J. L.; Bond, D. R.; Aksan, A.; Scriven, L. E. *Biotechnol. Prog.* **2007**, *23*, 2-17.
2. McDaniel, C. S.; McDaniel, J.; Wales, M.E.; Wild, J.R. In *Smart Coatings 2006*; Provder, T., Ed.; American Chemical Society Symposium Series XYZ; American Chemical Society: Washington, DC, 2007 pp. xx-yy.
3. Wang, P.; Sergeeva, M. V.; Lim, L.; Dordick, J. S. *Nature Biotechnol.* **1997**, *15*, 789-793.
4. Dordick, J. S.; Wang, P.; Sergeeva, M.V.; Novick, S.J. U.S. Patent 5,914,367, 1999.
5. Novick, S. J.; Dordick, J. S. *Biotechnol. Bioeng.* **2000**, *68*, 665-671.
6. Gill, I.; Ballesteros, A. TIBTECH **2000a**, *18*, 282-296.
7. Kim, Y. D.; Dordick, J. S.; Clark, D. S. *Biotechnol. Bioeng.* **2001**, *72*, 475-482.
8. Zhu, G.; Wang, P. *J. Am. Chem. Soc.* **2004**, *126*, 11132-11133.
9. Wang, L.; Zhu, G.; Wang, P.; Newby, B. M. S. *Biotechnol. Prog.* **2005**, *21*, 1321-1328.
10. El-Zahab, B.; Wang, P. *Biotechnol. Bioeng.* **2004**, *87*, 178-183.
11. Schmidt, A.; Dordick, J. S.; Hauer, B.; Kiener, A.; Wubbolts, M.; Witholt, B. *Nature* **2001**, *409*, 258-268.
12. Thomas, S. M.; DiCosimo, R.; Nagarajan, V. *Trends Biotechnol.* **2002**, *20*, 238-242.
13. Flickinger, M. C.; Freeman, A.E.; Anderson, C.R.; Lyngberg, O.K.; Laudon, M. C.; Scriven, L. E. Proceedings of *The Power of Ink-Jet Materials III*, **2005**, Vincenz Network, Hannover, GR.
14. Brányik,, T.; Kuncová, G.; Páca, J.; Demnerová, K. *J Sol-Gel Sci Technol* **1998**, *13*, 283-287.
15. Finnie, K. S.; Bartlett, J. R.; Woolfrey, J. L. *J. Mater. Chem.* **2000**, *10*, 1099-1101.
16. Gill, I.; Ballesteros, A. *Biotechnol. Bioeng.* **2000**, *70*, 400-410.
17. Fennouh, S.; Guyon, S.; Livage, J.; Roux, C. *J. Sol-Gel Sci.* **2000**, *19*, 647-649.
18. Livage, J.: Coradin, T.; Roux, C. *J. Phys. Condens. Matter,* **2001**, *13*, R673-R691.
19. Ferrer, M. L.; Yuste, Y.; Rojo, F.; del Monte, F. *Chem. Mater.* **2003**, *15*, 3614-3618.
20. Böttcher, H.; Soltzmann, U.; Mertig, M.; Pompe, W. *J. Mater. Chem.* **2004**, *14*, 2176-2188.
21. Fiedler, D.; Thron, A.; Soltmann, U.; Böttcher, H. *Chem. Mater.* **2004**, *16*, 3040-3044.
22. Buchholz, K; Kasche, V.; Bornscheuer, U. T. *Biocatalysis and Enzyme Technology,* Wiley-VCH Verlag GmbH, Weinheim, Germany, 2005.
23. Webb, C.; Dervakos, G. A. *Studies in Viable Cell Immobilization,* Academic Press, RG Landes: Austin, TX, 1996.

24. Karel, S. F.; Robertson, C. R. *Biotechnol. Bioeng.* **1989**, *34*, 320-336.
25. Karel, S. F.; Robertson, C.R. *Biotechnol. Bioeng.* **1989**, *34*, 337-356.
26. *Physiology of Immobilized Cells* ; De Bont, J. A. M.; Visser, J.; Mattiassen, B.; Tramper, J. Eds.; Elsevier: Amsterdam, The Netherlands, 1990.
27. *Immobilized Living Cells, Modeling and Experimental Methods;* Willert, R. G.; Baron, G. V.; De Backer, L. Eds.; John Wiley & Sons: Chichester, UK, 1996.
28. *Immobilization of Enzymes and Cells*; Guisan, J. M.; Ed.; Humana Press: Totowa, NJ, 2006.
29. *Immobilized Cells*; Wijffels, R. H. Ed.; Springer-Verlag: Berlin, Germany, 2001.
30. Shirey, T. *Clin. Chem.* **1981**, *16*, 147-155.
31. Walter, B. *Anal. Chem.* **1983**, *55*, 489A-514A.
32. Lawton, C. W.; Klei, H. E.; Sundstrom, D. W.; Voronoko, P. J. *Biotechnol. Bioeng. Symp.* **1986**, *17*, 507-517.
33. Lawton, C. W. M.S. thesis, University of Connecticut, Storrs, CT, 2001.
34. Bunning, T. J. M.S. thesis, University of Connecticut. Storrs, CT, 1988.
35. Schaffer, J. R.; Burdick, B. A.; Abrams, C. T. *CHEMTECH* September **1988**, 546-550.
36. Flanagan, W. P.; Klei, H. E.; Sundstrom, D. W.; Lawton, C.W. *Biotechnol. Bioeng.* **1990**, *36*, 608-616.
37. Cantwell, J. B.; Mills, P. D. A.; Jones, E.; Stewart, R. F. European Patent EP0288203, 1988.
38. Martens, N.; Hall, E. A. H. *Anal. Chimica. Acta.* **1994**, *292*, 49-63.
39. Swope, K. L.; Flickinger, M. C. *Biotechnol. Bioeng.* **1996**, *51*, 360-370.
40. Swope, K. L. Ph.D. thesis, University of Minnesota, Minneaplis, MN, 1995.
41. Swope, K. L.; Flickinger, M. C. *Biotechnol. Bioeng.* **1996**, *52*, 340-356.
42. Swope, K. L.; Liu, J.; Scriven, L. E.; Schottel, J. L.; Flickinger, M. C. In *Immobilized Cells: Basics and Applications;* Wijffels, R. H.; Buitelaar, B. M.; Burke, C.; Tramper, J. Eds.; Elsevier: Amsterdam, The Netherlands, 1996; pp 313-319
43. Lyngberg, O. K. Ph.D. thesis, University of Minesota, Minneapolis, MN, 2000.
44. Lyngberg, O. K.; Ng, C. P.; Thiagarajan, V. S.; Scriven, L. E.; Flickinger, M. C. *Biotechnol. Prog.* **2001**, *17*, 1169-1179.
45. Lyngberg, O. K.; Thiagarajan, V.; Stemke, D. J.; Schottel, J. L.; Scriven, L. E.; Flickinger, M. C. *Biotechnol. Bioeng.* **1999**, *62*, 44-55.
46. Lyngberg, O. K.; Stemke, D. J.; Schottel, J. L.; Flickinger, M. C. *J. Ind. Microbiol. Biotechnol.* **1999**, *23*, 668-676.
47. Huang, Z.; Thiagarajan,V. S.; Lyngberg, O. K.; Scriven, L. E.; Flickinger, M. C. *J. Colloid. Interface. Sci.* **1999**, *215*, 226-243.
48. Thiagarajan, V. S.; Huang, Z.; Scriven, L. E.; Schottel, J. L.; Flickinger, M. C. *J. Colloid. Interface. Sci.* **1999**, *215*, 244-257.
49. Ma, Y. Ph.D. thesis, University of Minnesota, Minneapolis, MN, 2002.

50. Lyngberg, O. K.; Solheid, C.; Charaniya, S.; Ma, Y.; Thiagarajan, V.; Scriven, L. E.; Flickinger, M. C. *Extremophiles* **2005**, *9*, 197-207.
51. Gebhard, M. S.; Lesko, P. M.; Brown, A. B.; Young, D.H. U. S. Patent 6750050, 2005.
52. Solheid, C. M.S. thesis, University of Minnesota, St. Paul, MN, 2003.
53. Fidaleo, M.; Charaniya, S.; Solheid, C.; Diel, U.; Laudon, M.; Ge, H.; Scriven, L. E.; Flickinger, M. C. *Biotechnol. Bioeng.* **2006**, *95*, 446-458.
54. Jons, S.; Ries, P., McDonald, C. *J. Membr. Sci.* **1999**, *155*, 79-99.
55. Desmone, M. F.; De Marzi, M.C.; Copello, G.J.; Fernández, M. M.; Malchiode, E.L.; Diaz, L.E. *Appl. Microbiol. Biotechnol.* **2005**, *68*, 747-752.
56. Gosse, J. L.; Engel, B. l.; Rey, F.; Harwood, C. S.; Scriven, L. E.; Flickinger, M. C. *Biotechnol. Prog.* **2007**, *23*, 124-130.
57. Charaniya, S. M.S. thesis, University of Minnesota. St. Paul, MN 2004.
58. Flickinger, M. C.; Mullick, A.; Ollis, D. F. *Biotechnol. Prog.* **1998**, *14*, 664-666.
59. Flickinger, M. C.; Mullick, A.; Ollis, D. F. *Biotechnol. Prog.* **1999**, *15*, 383-390.
60. Cussler, E. L. *Diffusion: Mass Transfer in Fluid Systems*, 2ed edn. Cambridge University Press, Cambridge, UK 1998.
61. Thiagarajan, V. S. M.S. thesis, University of Minnesota. St. Paul, MN 1998.
62. Thiagarajan, V. S.; Ming, Y.; Scriven, L. E.; Flickinger, M. C. In *Immobilized Cells: Basics and Applications;* Wijffels, R. H.; Buitelaar, B. M.; Burke, C.; Tramper, J. Eds.; Elsevier: Amsterdam, The Netherlands, 1996; pp 298-303.
63. Ge, H.; Zhao, C.-L.; Porzio, S.; Zhuo, L.; Davis, H. T.; Scriven, L. E.; In *Film Formation, Process and Morphology;* Provder, T., Ed.; American Chemical Society Symposium Series 941; American Chemical Society: Washington, DC, 2006; pp 69-90.
64. Gupta, A.; Singh, V. K.; Qazi, G. N.; Kumar, A. *J. Mol. Microbiol. Biotechnol.* **2001**, *3*, 445-456.
65. Mesa, M. M.; Caro, I.; Cantero, D. *Biotechnol. Prog.* **1996**, *12*, 709-712.
66. Diel, U. Diploma thesis, Technischen Fachochschule, Berlin, Germany, 2002.
67. Müh, T.; Bratz, e.; Ruckel, M. *Bioproc. Eng.* **1999**, *20*, 405-412.
68. Liu, J. M.S. thesis, University of Minnesota, St. Paul, MN, 1998.
69. Selifonova, O.; Burlage, R.; Barkay, T. *Appl. Environ. Microbiol.* **1993**, *60*, 3083-3090.
70. Schottel, J. L.; Orwin, P. M.; Anderson, C. R.; Flickinger, M. C. *unpublished.*
71. Van Dyk, T. K.; Wei, Y.; Hanafey, M. K.; Dolan, M.; Reeve, M. J. G.; Rafalski, J. A.; Rothman-Denes, L. B.; LaRossa, R. A. *Proc. Natl. Acad. Sci. USA* **2002**, *98*, 2555.

72. Jannek, K. Diploma thesis, Technical University Berlin, Berlin, Germany, 2006.
73. Oda, Y.; Samanta, S. K.; Rey, F. E.; Wu, L.; Liu, X.; Yan, T.; Zhou, J.; Harwood, C. S. *J. Bacteriol.* **2005**, *187*, 7784-7794.
74. Rey, F. R.; Oda, Y.; Harwood, C.S. *J. Bacteriol.* **2006**, *188*, 6143-6152.
75. Bond, D. R.; Holmes, D. E.; Tender, L. M.; Lovely, D. R. *Science* **2002**, *295*, 483-485.
76. Bond, D. R.; Lovely, D. R. *Appl. Environ. Microbiol.* **2003**, *69*, 1548-1555.
77. Gregory, K. B.; Bond, D. R.; Lovely, D. R. *Environ. Microbiol.* **2004**, *6*, 596-604.
78. Reguera, G. K.; McCarthy, D.; Mehta, T.;Nicoll, J. S.; Tuominen, M. T.; Lovely, D. R. *Nature* **2005**, *435*, 1098-1101.
79. Srikanth, S., Marsili, E., Flickinger, M. C., Bond, D. R. *unpublished.*
80. Fuller, B. J.; Lane, A. N. Eds.; *Life in the Frozen State;* CRC Press: Boca Raton, FL, 2004.
81. Webb, S. J. *Bound Water in Biological Integrity*, Charles Thomas Publisher: Springfield, IL, 1965.
82. Xie, G.; Timasheff, S. N. *Protein Sci.* **1997**, *6*, 211-221.
83. Potts, M. *Microbiol. Rev.* **1994**, *58*, 755-805.
84. Carley, S.; Record, M.T. *Biochemistry* **2003**, *42*, 12596-12609.
85. Felitsky, D. J.; Cannon, J. G.; Capp, M. W.; Hong, J.; Van Wynsberghe, A. W.; Anderson, C.F.; Record, M.T. *Biochemistry* **2004**, *43*, 14732-14743.
86. Attfield, P. V. *FEBS Lett* **1987**, *225*, 259-263.
87. Scheidler, P.; Kob, W. L. *Europhysics Lett.* **2002**, *59*, 701-707.
88. Schreiber, A., Ketelsen, I.; Findenegg, G. H. (2001). *Phys. Chem. Chem. Phys.* **2001**, *3*, 1185-1195.
89. Hansen, E. W., Stocker, M.; Schmidt, R. (1996). *J. Phys. Chem.* **1996**, *100*, 2195-2200.
90. Zhang, J.; Liu, G. *J. Physic. Chem.* **1992**, *96*, 3478-3480.
91. Ellison, C. J.; Torkelson, J. M. *Nature Materials* **2003**, *2*, 695-700.
92. Aksan A.; Irimia, D.; He, X.; Toner, M. *J. Appl. Phys.* **2006**, *99*, 064703.
93. Lyngberg, O. K.; Flickinger, M. C.; Scriven, L. E.; Anderson, C. R. U.S. Patent 7132247, 2006.

Chapter 4

Smart Surfaces for the Control of Bacterial Attachment and Biofilm Accumulation

Linnea K. Ista, Sergio Mendez, Sreelatha S. Balamurugan,
Subramanian Balamurugan, Venkata G. Rama Rao,
and Gabriel P. Lopez

Center for Biomedical Engineering, Department of Chemical and Nuclear
Engineering, University of New Mexico, Albuquerque, NM 87131

Controlling the nonspecific attachment of bacteria to surfaces is important for various human endeavors, from maintaining clean submerged surfaces to biosensing. Surface grafted stimuli-responsive polymers (SRPs) represent a unique group of biomaterials whose dynamic surface energetics and molecular topography can result in reversible attachment and release of microbial cells and proteins. Experiments using model solid surfaces composed of grafted poly(N-isopropylacrylamide) have demonstrated and reiterated the importance of hydrophobicity and hydrogen bonding on bacterial attachment and extended these principles to mechanisms of release; furthermore, both binding and release can be tuned by altering the surrounding chemical milieu of the grafted SRP. Recent studies with self assembled monolayers containing oligo(ethylene glycol) have demonstrated the importance of molecular rearrangement on attachment (or resistance thereto) and release of attached cells and proteins. Taken together, these studies indicate new approaches for not only mitigation of bacterial and protein accumulation, but directed and dynamic promotion of attachment and release.

Introduction

A defining characteristic of an important class of stimuli responsive polymers (SRPs), a rapid and reversible phase transition in response to a small, specific change in environmental conditions, translates to an interesting interfacial property when such polymers are grafted onto a surface. The collapse and expansion of these so called "smart polymers" through the transitions results in a change in water contact angle (Fig. 1).

This property of switchable surface energy has been exploited for a number of processes, including chromatography and drug delivery.[3] An early discovery by Okano and workers that surface grafted poly(N-isopropylacrylamide) (PNIPAAM), one of the most widely studied SRPs, could be used to release attached cultured cells[4] led us to explore the possibility that a similar process could be exploited to reversibly attach and release bacterial cells, and thus, control biofilm formation. Bacteria did, indeed, exhibit triggered release from surface-grafted PNIPAAM. Previously, we noted that different species of bacteria attach differently in response to surface hydrophobicity.[2,5] The preference for either a hydrophobic or a hydrophilic surface carried over to the

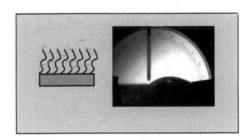

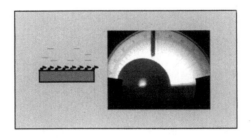

Figure 1. Wettability of plasma-initiated surface grafted PNIPAAM at 25°C (top) and 45°C (bottom). The change in water contact angle corresponds with a reversible collapse and expansion of the polymer as it goes through its transition temperature (T_t) of ~32°C.

release experiments as well; bacteria that attach well to more hydrophobic surfaces released best when attached above the transition temperature (T_t ~32°C for PNIPAAM) and rinsed with buffer below this temperature; other bacteria, which seemed to prefer hydrophilic surfaces, released only when attached below T_t and rinsed with buffer above it.[6]

Other researchers, notably Cunliffe,[7] have shown not only a difference in attachment above and below T_t, but have shown similar release (i.e. >90%) of bacteria on PNIPAAM. Furthermore, similar results were obtained for proteins under similar conditions.[7]

While the fouling release phenomenon has been firmly established, the molecular processes behind it remain largely unresolved. In this paper we describe some of ways different types of surface-grafted SRPs can be used to address this question and demonstrate their potential utility in a variety of applications.

Materials and Methods

Media and buffers

All media and buffers were prepared with deionized water (d-H$_2$O) generated by a system in which tap water was subjected sequentially to water softening, reverse osmosis, and ion exchange (Barnstead-Thermolyne RoPure/Nanopure system). The final resistivity of the processed water was greater than 18 MΩ cm^{-1}.

Preparation of plasma-initiated PNIPAAM [6]

N-isopropylacrylamide (NIPAAM) (Aldrich Chemicals) was recrystallized from hexane and dissolved in 2-propanol at 10mg mL^{-1}. This solution was degassed by repeated freeze-thaw cycles under a vacuum and was kept under argon until use. Polystyrene coupons were cut from the lids of standard laboratory petri dishes (VWR) soaked in ethanol, rinsed in d-H$_2$O, and dried under nitrogen. These coupons were then subjected to an argon plasma treatment (pressure, 75 μm Hg) for 5 min with in a Harrick model RDC-32G plasma cleaner-sterilizer. Exposure of the polystyrene samples to the argon plasma resulted in the formation of free radicals on the surfaces of the polystyrene. The free radicals initiated polymerization upon addition of the NIPAAM monomer solution (under argon) to the plasma-treated polystyrene in the plasma cleaner and sealing of the system. Polymerization was allowed to continue for 1 h, after which the NIPAAM solution was removed. The samples were then rinsed with three 25 mL washes of 2-propanol to remove the residual monomer followed by dH$_2$O; they were dried with a stream of dry N$_2$. Samples were stored in the dark until they were used (usually no more than 1 week). Polystyrene samples that were plasma treated but were not used for grafting were kept as control samples and were designated plasma-cleaned polystyrene (PCPS).

Preparation of self-assembled monolayers (SAMs) of ω-terminated alkane thiolates on gold [2]

SAMs of ω-terminated alkanethiolates were used as controls for attachment studies, substrata for *in situ* polymerization procedures and as fouling resistant/release coatings themselves. SAMs were prepared on gold films evaporated onto glass microscope slides or coverlips (VWR Scientific). The glass substrata were cleaned by immersion in a solution (piranha etch) prepared by mixing 70% (vol/vol) concentrated H$_2$SO$_4$ with 30% H$_2$O$_2$ for 20 min to 1 h, thoroughly rinsing the slides in dH$_2$O and drying them under a stream of dry N$_2$. *Piranha etch is a powerful oxidizer and can react violently when placed in contact with organics and should be stored only in containers which prevent pressure buildup.* The samples were then placed into the vacuum chamber of a metal evaporator. The system was evacuated to 10^{-7} torr, and 10Å of chromium followed by 300 Å of gold were deposited on the substrata. The system was restored to room pressure, and the samples were removed and submerged in 1 mM ethanolic solutions of dodecanethiol (referred to herein as CH$_3$-thiol and obtained from Aldrich Chemical), 11-mercapto-1-undecanol (OH-thiol; Aldrich Chemical), 2-mercaptodecanoic acid (COOH-thiol; Aldrich Chemical), 1-hexa(ethylene glycol) undecanethiol (OEG-thiol and obtained from Prochimia Gdansk, Poland)or mixtures of two of these thiols. The samples were immersed in thiol solution overnight at 4°C, after which they were rinsed in ethanol and dried under a stream of dry N$_2$. The resulting surface (i.e., the SAMs of ω-terminated alkanethiolates) will be referred to as CH$_3$-SAM, OH-SAM, COOH-SAM, OEG-SAM or, in the case of mixed monolayers, COOH/CH$_3$-SAM, OH/CH$_3$-SAM, OEG/CH$_3$ SAM.

ATRP production of copolymer surfaces

N-Isopropyl acrylamide/*tert*-butyl acrylamide copolymers were grafted from mixed SAMs on gold by atom transfer radical polymerization (ATRP) using a CuBr and Me$_4$Cyclam (1,4,8,11-tetramethyl 1,4,8,11-azacyclo-tetradecane) catalyst system at room temperature.[8] All the chemicals were purchased from Aldrich Chemical Co., Inc. The monomers were recrystallized from hexane. As a substrate for graft polymerization, we used OH/CH$_3$ SAMs with a surface mole fraction of hydroxyl-terminated alkylthiolates ~0.32, as estimated previously.[8] To perform ATRP, an initiator was immobilized on the SAM surface by reacting the hydroxyl groups with 2-bromopropionyl bromide in the presence of triethylamine in THF (anhydrous). Polymerization was carried out by immersing the substrate in a solution containing NIPAAM and TBA (2M total; varying proportions), 0.2mM CuBr, and 0.2mM Me$_4$Cyclam in DMF (anhydrous) at room temperature for 1 h. Polymerization was done inside a glovebox. After polymerization, the samples were washed with DMF, ethanol, and d-H$_2$O.

Tunable PNIPAAM

PNIPAAM was grafted from initiator-modified CH$_3$/COOH- SAMs of varying mole fractions.[9] The COOH moieties were activated by incubation in 10 mM 2-ethyl-5-phenylisoxazolium-3-sulfonate (Woodward's Reagent K, Aldrich) aqueous solution for 30 min. After being rinsed with deionized water, the samples were immersed in 200mM of the free radical initiator, 2,2-azobis(2-amidinopropane) hydrochloride ("ABAH", DuPont) aqueous solution for 2 h. These samples were immediately placed in a degassed NIPAAM aqueous solution (~25% w/v). The polymerization was performed at 65 °C for 3 h. Extensive rinsing of the samples with d-H$_2$O was carried out after this step.

Silica-PNIPAAM composites[10]

Silica sol was prepared by a standard sol-gel process by mixing tetraethyl orthosilicate (TEOS, Aldrich), ethanol, water, and HCl in a molar ratio of 1:3:1:0.0007 followed by a reaction at 60 °C for 90 min. The resulting stock sol was stored at -20 °C until use. For each membrane prepared, 0.25 mL of stock sol was diluted with 0.043 mL of water and 0.6 mL of ethanol and stirred well to yield a sol with final proportion (TEOS:ethanol:water:HCl) 1:20:5.02:0.0007. Aliquots (1 mL) of PNIPAAM (Mw= 349,000; Mw/Mn ~3.64, Polysciences, Inc) solutions (4.5 mg/mL) in pH 4.0 phosphate buffer were prepared, added to the diluted stock sol, and stirred to obtain a clear, transparent hybrid sol. The concentration of PNIPAAM in precursor solution were varied in vol % to that of silica. The hybrid sol was spin coated onto piranha-cleaned glass slides and allowed to dry before exposure to bacterial suspension.

Bacterial strains and culture conditions

Marine broth 2216 (MB, Difco) was prepared according to the manufacturer's instructions. Marine agar (MA) was prepared by the addition of 1.5% Bacto agar (Difco) to MB. Artificial sea water (ASW) contained 400 mM NaCl, 100 mM $MgSO_4$, 20 mM KCl, and 10 mM $CaCl_2$.[11] Modified basic marine medium plus glycerol (MBMMG) contained 0.5× ASW plus 19 mM NH_4Cl, 0.33 mM K_2HPO_4, 0.1 mM $FeSO_4 \cdot 7H_2O$, 5 mM Trishydroxyaminomethane hydrochloride pH 7, and 2 mM glycerol.[6, 11] Nutrient broth (NB, Difco), was prepared according to the manufacturer's directions. Nutrient agar (NA) was made by the addition of 1.5 % Bacto agar to NB. *Staphylococcus* basal medium plus glycerol (SBMG) contains 6 mM $(NH_4)SO_4$, 0.5 mM $MgSO_4 \cdot 7H_2O$, 13.5 mM KCl, 28 mM KH_2PO_4, 72 mM Na_2HPO_4, 1 μg mL^{-1} thiamin; 0.5 μg mL^{-1} biotin; 0.5% Bacto-Peptamin (Difco) and 1mM glycerol.[2, 6] Delbecco's phosphate buffered saline (PBS; Sigma Chemical) was prepared according to manufacturer's instructions.

Cobetia marina (basonym, *Halomonas marina*)

ATCC 25374 was revived from the original lyophilate and was stored as frozen stock aliquots in MB +20% glycerol at -70°C. Experimental stock cultures were maintained on MA slants and were stored at 4°C for up to 2 weeks. Prior to inoculation into a chemostat, a single colony from a MB slant was inoculated into 50 mL of MB and grown overnight with shaking at 25°C. A chemostat culture was established by inoculating 3 mL of the overnight culture into MBMMG.[2, 11] The chemostat was maintained at a flow rate of 1 mL min^{-1} (dilution rate, 0.16 h^{-1}) with constant stirring. The concentration of the chemostat culture was 10^7 cells mL^{-1}.

Staphylococcus epidermidis

ATCC 14990 was revived in NB from the original lyophilate. Aliquots were stored in NB + 20% glycerol at -70°C. Experimental stock cultures were maintained on NA slants and were stored at 4°C for up to 2 weeks. Prior to inoculation into a chemostat, a single colony was transferred into 50 mL of NB and grown overnight at 37°C with shaking. Five mL of this culture was inoculated into 500 SBMG. The flow rate was 1.5 mL min^{-1}, and the dilution rate was 0.16 h^{-1}. The bacterial concentration was ~5 x10^7 cells mL^{-1}.

Attachment and detachment studies [6,12]

Attachment/detachment experiments used surface grafted PNIPAAM samples and controls placed in sterile glass Petri dishes containing either (i) ASW containing 5x 10^6 C. *marina* cells mL^{-1}, or (ii) PBS containing 10^7 S. *epidermidis* cells mL^{-1} preequilibrated for 20 min. at either 37°C or 25°C respectively. After incubation at the preequilibration temperature for 2 h, the samples were rinsed with d-H_2O at the same temperature, dried with a stream of N_2, and examined with a phase-contrast microscope (Nikon Labophot). Images of 10 randomly chosen fields of view were captured with a 60x objective by using a digital camera interfaced with Image Tool image-processing software. Image capture and analysis were done as previously described.[6] The samples were then washed in 60 mL either 4° C ASW (for C. *marina*) or 37°C PBS (for S. *epidermidis*) delivered from a syringe, rinsed in d-H_2O, and recounted.

Results and Discussion

Release of bacteria from plasma- polymerized PNIPAAM surfaces

Figure 2 demonstrates the degree to which two bacteria, *Cobetia marina* and *Staphylococcus epidermidis,* are released from PNIPAAM surfaces as compared to controls, after both short term (i.e 2hr.) and long term (i.e., 3 day) incubations at a favorable attachment temperature. These two strains are known to attach quite differently in response to substratum wettability: C. *marina* attaches most strongly to hydrophobic surfaces, while S. *epdidermidis* attaches much more strongly to hydrophilic surfaces. As such, their attachment and detachment on PNIPAAM are quite different: C. *marina* is only released when attached above T_t (i.e. when the polymer is collapsed and relatively hydrophobic) and the temperature lowered below T_t whereas S. *epidermidis* is only detached when attached below T_t (i.e. the polymer is extended and relatively hydrophilic) and the temperature raised above T_t.

Although the film in this study was quite thin, the polymer was quite effective in removing biofilms.[6] The relative ease with which plasma initiated films can be synthesized make them attractive for studying bacterial surface interactions. Recent work on mammalian cells also suggests that such films can be used to examine the cellular components that remain after detachment, an important considertation when developing non-fouling surfaces.[13] One disadvantage of such films, however, is the uncertainty of the chemistry of the underlying substratum, which may also contribute to (or inhibit) fouling release.

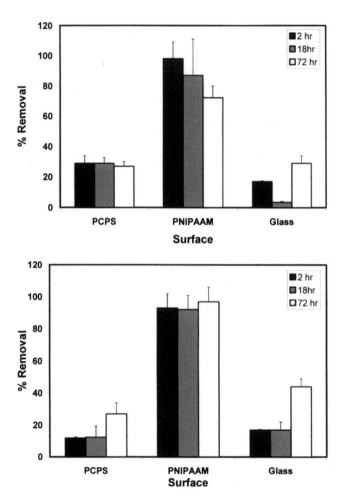

Figure 2. Percent detachment of Staphylococcus epidermidis *(top) and* Cobetia marina *(bottom) from plasma-grafted PNIPAAm at 2, 18 and 72 hours.* S. epidermidis *was incubated at 25°C for the specified time, and the number of cells counted before and after rinsing with 60 mL phosphate buffered saline at 37°C.* C. marina *was incubated with the samples at 37°C for the specified amount of time and the number of cells counted before and after rinsing with 60 mL artificial seawater. Controls were plasma cleaned polystyrene (PCPS) and piranha cleaned glass (Glass).[2]*

Surfaces with well-defined underlying chemistry can be generated by grafting PNIPAAM from initiator modified self-assembled monolayers (SAMs) of ω-terminated alkanethiolates. Our group has used two protocols for this process: the first, used a convential free-radical initiator grafted onto a COOH-SAM.[12] While this procedure produced PNIPAAM surfaces that functioned effectively in reversibly attaching bacteria, the synthesis procedure was cumbersome, and the reproducibility unreliable. A second method, utilizing atom transfer radical polymerization (ATRP) was developed in which the radical itself is confined to the growing polymer chain, thus eliminating early termination and contamination of surfaces by free radicals that initiate polymerization in solution.[8] In addition, since the polymerization in this procedure can be done in nonaqueous solutions, incorporation of other comonomers, which may not be as water soluble is possible. Below we discuss two applications of PNIPAAM grafted from SAMs.

While the properties of PNIPAAM make it excellent for use in medical applications, where a T_t of 32°C is practical and attainable, environmental uses of PNIPAAM for biofilm control (e.g. on ships, power plant heat-exchangers, oil platforms) require different transition temperatures. For this reason, we have investigated the use of copolymers of NIPAAM and *tert*-butylacrylamide (TBA) for fouling release. As changing the hydrophobicity of the monomers using this method results in a change in the T_t, use of such polymers also allows for the examination of this change on bacterial interaction with the surfacs. We made two copolymers using ATRP polymerization from hydroxyl-terminated alkanethiolates: 20 and 40 wt% TBA were used in the final polymerization step, yielding surfaces with T_t of 21 and 14 degrees respectively, which will be subsequently denoted as P21 and P14 respectively. The advancing water contact angles for P21 were 78+/-2° above T_t and 63+/- 2° below; those for P14 were 81 +/- 2° above and 60+/- 2° below. P21 is similar to the polymer used by Cunliffe.[7] Figure 3 demonstrates the release of bacteria from copolymer surfaces.

Figure 3A presents attachment and release data for the copolymers obtained using our standard assay for *C. marina* on PNIPAAM: attachment of the organism for two hours at 37°C, counting of attached cells and release under shear at 4°C. While the general trends noted for *C. marina* with regard to bacterial attachment are observed (i.e. more cells attach to the more hydrophobic P14 than to P21), the efficiency of release is much lower that observed previously on the surface-grafted homopolymer. As seen below, *C. marina* seems to detach best when the surface goes from moderately hydrophobic to moderately hydrophilic. Figure 3B demonstrates attachment and release when the experiment is repeated with attachment occurring at 25°C and release at 4°C. While the contact angle change is the same, the percentage of cells decreases dramatically. In addition, the number of cells attaching to polymers (relative to the methyl-terminated SAM control) increased. Analysis of this result implies that the change in incubation temperature, while not changing the global

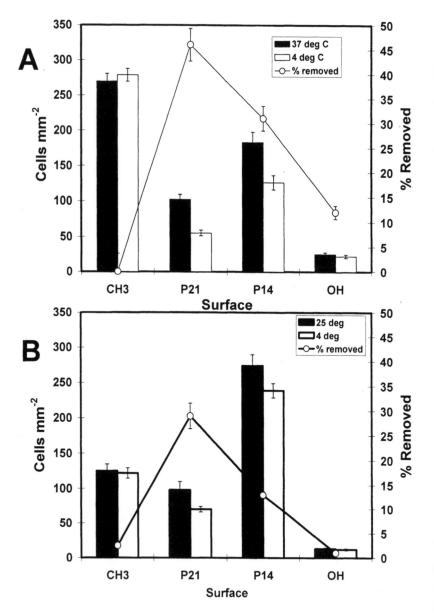

Figure 3. Attachment and detachment of Cobetia marina *on NIPAAM/TBA copolymer surfaces. A. Attachment for two hours at 37°C, B. Attachment for two hours at 25°C*

properties of the polymers, may have a profound effect on bacterial physiology as it applies to attachment and release. Therefore, caution must be exercised when comparing attachment and release at two different temperatures.

When a series of PNIPAAM surfaces were prepared from initiator modified SAMs with different surface mole fractions of initiator moiety, it was observed that, for very thin (i.e. <250 Å) films, the composition and surface properties of the underlying SAM changed the contact angles of the overlying PNIPAAM film.[9] Use of such films allows for variation in the contact angles above and below T_t while keeping T_t itself constant, and, thus, allowing for direct comparison of different contact angle ranges. Using tunable surfaces to study bacterial attachment and detachment, we observed quite different behavior between the two strains in question, which may be related to the nature of the transition used for each strain. Figure 4 presents the attachment data for bacteria on these surfaces. Both species attached as expected: i.e. more *C. marina* attached as the contact angle of the surfaces increased, while *S. epidermidis* attached in greater numbers to surfaces with decreasing contact angle.

Detachment however, was quite different between the strains. While *S. epidermidis* detachment correlated strongly with the contact angle of the polymer under both attachment and detachment temperatures (Fig. 5), detachment of *C. marina* was greatest at intermediate contact angles.

The reasons behind these differences in detachment of organisms when going from lower to higher temperature and those going from higher to lower, as well as the reasons for the tunable phenomenon itself remain unclear. The process of hydration is quite different than the process of rehydration and bacteria attaching to the hydrated state may do so through different molecular interactions than those attaching to PNIPAAM in its collapsed state. In addition, the role of surface density of the initiator, and the resultant polymer chains, may play a role. It is known, for example, that grafted PNIPAAM exhibits the greatest freedom of movement during the transition when the surface density of the initiator is ~0.6[14] and initial results indicate that this surface density also promotes the greatest release of *C. marina* (data not shown).

Combining copolymerization and tunable PNIPAAM surfaces will greatly improve the utility of these surfaces as it will allow for a range of temperatures and hydrophobicities to be tailored for a specific application. A second means by which SRP films can be made more practical for applications is incorporation of the "soft" polymers into a more durable matrix that allows coating of a variety of substrates. We previously demonstrated the utility of such an architecture using a silica matrix for the creation of molecular valves.[10] We explored this methodology to fouling release. Figure 6 demonstrates the attachment and detachment characteristics of different weight percents of PNIPAAM in a silica gel matrix. Surprisingly, the hybrid films attached fewer bacteria than either the methyl-terminated SAM control or the pure silica films. In addition, for the best performing hybrid films, approximately 80% of attached cells were released. The result is a hybrid film that combines fouling resistance

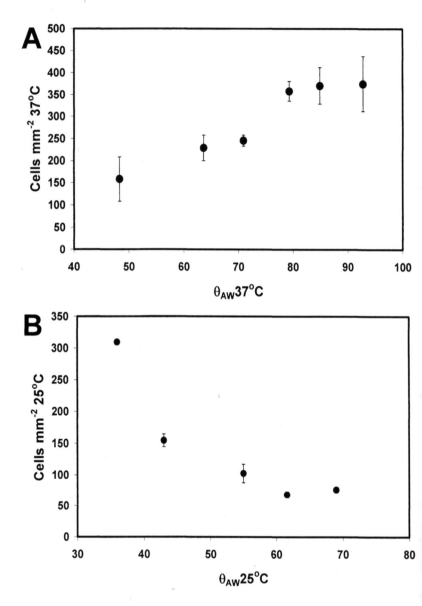

Figure 4. Attachment of C. marina *(A) at 37°C and* S. epidermidis *(B) at 25°C to tunable SAMs as a function of* θ_{AW} *at the attachment temperature.*

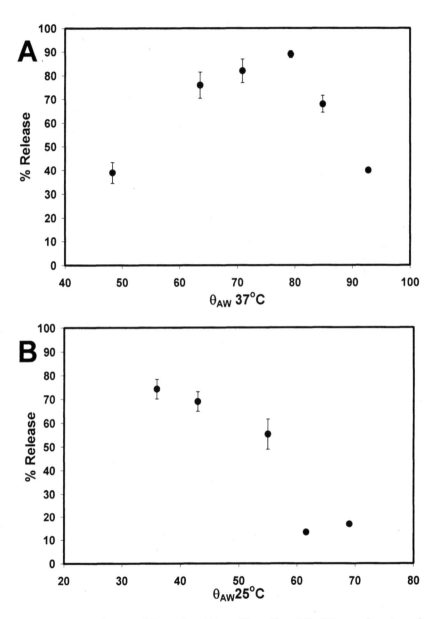

Figure 5. Detachment of C. marina *(A) and* S. epidermidis *(B) as a function of the advancing water contact angle of tunable PNIPAAM at the attachment temperature.*

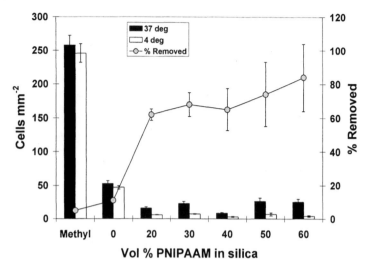

Figure 6. Attachment and release of C. marina *on silica-PNIPAAm surfaces with varying vol% of PNIPAAm. Controls are a methyl-terminated alkanethiolate on gold (methyl) and pure silica (0).*

with fouling release. We are currently exploring the nature of both the fouling resistance and release, in addition to fine-tuning the coatings for maximum performance.

That fouling resistant surfaces can also exhibit phase-dependent fouling release is not surprising. Early work by Prime, et al.[15] on the protein resistance of oligo (ethylene glycol) terminated alkanethiolate SAMs suggested that mixed monolayers of OEG and methyl were capable of reversibly adsorbing and desorbing proteins upon a change in temperature. We have extended this work[1] and shown that the protein release correlates with a small, reversible change in advancing water contact angle at ~30°C. *C. marina* was also shown to attach above this T_t, and to detach upon rinsing at 4°C (Fig. 7). The release was substantial given the contact angle change between high and lower temperature (~7°), and seems to be greater than expected from a mere change in hydrophobicity alone. Analysis of the mixed SAMs with sum frequency generation spectroscopy indicates that both protein and bacterial release may be due to hydration of oligo (ethylene glycol chains) at lower temperature, which may restore fouling resistance properties to the SAMs.[1]

Acknowledgements

This work is supported by grants from the Office of Naval Research.

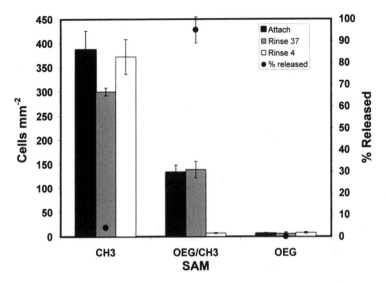

Figure 7. Attachment and release of C. marina *on pure CH₃ and OEG-SAMs and OEG/CH₃ SAMs with a surface mole fraction OEG of 0.5. Cells were attached for two hours at 37°C, counted and rinsed with 60 mL ASW at either 37°C or 4°C. % release upon rinsing at 4°C is depicted on the second y-axis.* [1]

Literature Cited

1. Balamurugan, S.; Ista, L. K.; Yan, J.; Lopez, G. P.; Fick, J.; Himmelhaus, M.; Grunze, M., Reversible protein adsorption and bioadhesion on monolayers terminated with mixtures of oligo(ethylene glycol) and methyl groups *J. Amer. Chem. Soc.* **2005**, *127*, (42), 14548-14549.

2. Ista, L. K.; Fan, H.; Baca, O.; López, G. P., Attachment of bacteria to model solid surfaces: oligo(ethylene glycol) surfaces inhibit bacterial attachment. *FEMS Microb. Lett.* **1996**, *142*, 59-63.

3. Hoffman, A. S., Environmentally Sensitive Polymers and Hydrogels. *MRS Bulletin* **1991**, *16*, 42-46.

4. Okano, T., *et al*, A Novel Recovery System for Cultured Cells Using Plasma-Treated Polystyrene Dishes Grafted with Poly(*N*-isopropyl-acrylamide). *Journal of Biomedical Materials Research* **1993**, *27*, 1243-1251.

5. Ista, L. K.; Callow, M. E.; Finlay, J. A.; Coleman, S. E.; Nolasco, A. C.; Callow, J. A.; Lopez, G. P., Effect of substratum surface chemistry and surface energy on attachment of marine bacteria and algal spores. *Appl. Environ. Microbiol.* **2004**, *70*, (7), 4151-4157.

110

6. Ista, L. K.; Pérez-Luna, V. H.; López, G. P., Surface-grafted, enivironmentally responsive polymers for biofilm release. *Appl. Environ. Microbiol.* **1999**, *65*, (4), 1603-1609.

7. Cunliffe, D.; de las Heras Alarcon, C.; Peters, V.; Smith, J. R.; Alexander, A., Thermoresponsive surface-grafted poly(*N*-isopropylacrylamide) copolymers:effect of phase transitions on protein and bacterial attachment. *Langmuir* **2003**, *19*, (7), 2888-2899.

8. Balamurugan, S.; Mendez, S.; Balamurugan, S. S.; O'Brien, M. J.; Lopez, G. P., Thermal response of poly(N-isopropylacrylamide) brushes probed by surface plasmon resonance *Langmuir* **2003**, *19*, (7), 2545-2549.

9. Mendez, S.; Ista, L. K.; Lopez, G. P., Tuning wettability with smart polymers grafted on mixed self-assmbled monolayers. *Langmuir* **2003**, *19*, (19), 8115-8116.

10. Rama Rao, G. V.; Krug, M. E.; Balamurugan, S.; Xu, H. F.; Xu, Q.; Lopez, G. P., Synthesis and characterization of silica-poly(N-isopropylacrylamide) hybrid membranes: Switchable molecular filters *Chemistry of Materials* **2002**, *14*, (12), 5075-5080.

11. Kersters, K., The genus *Deleya*. In *The Prokaryotes*, 2 ed.; Balows, A.; Trüper, H. G.; Dworkin, M.; Harder, W.; Schliefer, K. H., Eds. Springer-Verlag: New York, 1992; Vol. 4, pp 3189-3197.

12. Ista, L. K.; Mendez, S.; Perez-Luna, V. H.; Lopez, G. P., Synthesis of poly(N-isopropylacrylamide) on initiator-modified self-assembled monolayers. *Langmuir* **2001**, *17*, (9), 2552-2555.

13. Canavan, H. E.; Cheng, X. H.; Graham, D. J.; Castner, D. G.; Ratner, B. D., Cell sheet detachment affects the extracellular matrix: A surface science study comparing thermal liftoff, enzymatic, and mechanical methods. *Journal of Biomedical Materials Research* **2005**, *74A*, (1), 1-13.

14. Yim, H.; Kent, M. S.; Satija, S.; Mendez, S.; Balamurugan, S. S.; Balamurugan, S.; Lopez, G. P., Study of the conformational change of poly(N-isopropylacrylamide)-grafted chains in water with neutron reflection: Molecular weight dependence at high grafting density *Journal of Polymer Science B-Polymer Physics* **2004**, *42*, 3302-3310.

15. Prime, K. L.; Whitesides, G. M., Adsorption of proteins onto surfaces containing end-attached oligo(ethylene oxide): a model system using self-assembled monolayers. *J. Amer. Chem. Soc.* **1993**, *115*, 10714-10721.

Chapter 5

Performance Testing of Waterborne Antibacterial and High-Solids Coatings

D. L. Clemans[1], S. J. Rhoades[1], J. J. Kendzorski[1], Q. Xu[2], and J. Baghdachi[2]

[1]Department of Biology and [2]Coatings Research Institute, Eastern Michigan University, Ypsilanti, MI 48197

In a previous publication (1a), the formulation of the test model Organic Antibacterial Coating, and its effectiveness against *Escherichia coli ATCC* 11229 and *Staphylococcus aureus* ATCC 6538 was characterized. This paper aims to review our continued bacterial challenges of the coating, including the results produced upon introduction of various conditions the coating was subjected to, during and prior to the test, and extending the panel of bacteria used in the challenge to include *Bacillus subtilis* ATCC 6633 and *Pseudomonas aeruginosa* ATCC 15442. In previous studies we have demonstrated that air dry waterborne coatings containing silver-based antimicrobials effectively kill the test strains of *E. coli* and *S. aureus* at a concentration of 10^5 CFU/ml after 12 hours of incubation at RT. In this study, we have also demonstrated that these coatings are just as effective against *B. subtilis* and *P. aeruginosa* after 12 hours of incubation. Results when analyzing the coatings effectiveness before 12 hours showed no significant reductions before this time point for any of the test organisms. We have also shown that the coating is still effective when exposed to UV radiation, increased humidity, increased temperature, a common cleaning agent and varied organic loads. The substrate was also changed from glass to ceramic tiles and no loss in efficacy was observed. When increasing the bacterial load from 10^5 to 10^9 CFU/ml, the observed reductions were much less significant. After 24 hours at this bacterial concentration only 31% of *E. coli* and 72% of *S. aureus* were killed.

Introduction

Many infectious diseases can be acquired through casual contact with infected individuals and environmental sources. Routine sanitation by the use of disinfecting agents is fairly successful in combating the spread of some infectious diseases. However, cleaning procedures can be costly and time consuming, and do not offer continuous protection or guard against fresh contamination. Self-disinfecting surface coatings containing non-migratory antimicrobial agents can fulfill this task. The use of such surface coatings containing non-migratory antimicrobial agents, while not aimed at producing a totally sterile environment, may serve a useful purpose in areas where a permanent and more sanitary environment is necessary.

Coatings can be designed to kill bacteria through three mechanisms: 1) coatings that can resist the attachment of bacteria, 2) coatings that release biocides that will kill the bacteria, and 3) coatings that kill bacteria on contact (1-4). Coatings can also combine two or more of these mechanisms. The coatings used in these challenges fall into the second group, those that release biocides. The biocide used is the incorporation of silver ions into zeolites.

Zeolites are inorganic silicates that contain pores large enough to hold metal ions. Silver ions can be absorbed into zeolites and then released at a controlled rate. Coatings containing silver filled zeolites may be suitable for controlling and preventing the growth of a wide variety of microorganisms. One specific agent is a zirconium phosphate complex of silver ions which has found commercial application as an antimicrobial agent in textile, filtration media, building materials and adhesives.

Silver has a very long history of use as an effective, broad-spectrum antimicrobial agent (13-18). Silver was used as an antimicrobial agent at least as long ago as 1000 BC (19) in applications ranging from water treatment to medicine. Silver is used today to impart bactericidal properties to water filtration devices, as an alternative to halogen-based products in swimming pools, spas and small cooling water systems (20-23), and to control infections in hospitals (19, 21, 23). Clinical applications include Ag-sulphadiazine for treatment of burns (24), $AgNO_3$ for prevention of gonorrhea opthalmicum in neonates (23), and Ag-complexes for dental resin composites (25). The reasons for the historical use of silver as a biocide are based on the relative safety to mammals and efficacy against a broad range of microorganisms. Recent papers suggest that the mode of action of silver ions varies depending upon the target organism, and can interact with functional groups on proteins containing sulfur, phosphorus, nitrogen, and oxygen (26). Further, silver ions have been shown to disrupt the TCA cycle, electron transport system, phosphate uptake, transport of various metabolites, and the polarization of bacterial membranes (26).

The antimicrobial activities of these coatings on the test substrates are tested toward a selected panel of Gram-positive (*Staphylococcus aureus* ATCC 6538,

Bacillus subtilis ATCC6633) and Gram negative (*Escherichia coli* ATCC 11229, *Psuedomonas aeruginosa* ATCC15442) bacteria. These strains of bacteria have been used as standards for testing of disinfectant efficacy using the AOAC Use-Dilution Methods and represent common bacteria encountered in health care and food preparation facilities (35, 36, 37-40).

Materials and Methods

Preparation of antimicrobial coatings

Antimicrobial coatings are formulated using thermoset and thermoplastic resins, standard additives, and by incorporating various levels of various silver containing antimicrobial agents. The antimicrobial agents used were purchased from Milliken (A), and Ciba Specialty Products (B). The agents were further treated with various proprietary agents to impart certain properties and increase their compatibility with various resins and additives. The antimicrobial coating formulas include water-borne coatings and solvent-borne coatings. The resins used in the formulas of the antimicrobial coatings include acrylic thermoplastic solution (Chempol 317-0066, CCP), acrylic emulsion (Neocryl A622, Flexbond 381, Air Products), polyurethane dispersion (Neorez R9679, Neoresins Inc.), polyester dispersion Bayhydrol XP 7093 and Byhydur polyisocyanate, (Bayer Material Science) and urethane/ acrylic copolymer dispersion (Hybridur 570, Air Products).

In most cases, the coatings were applied onto glass, ceramic, and previously painted surfaces by a draw down method ensuring film thicknesses in the range of 20-50 micrometers.

Air-dry acrylic coating formulation (06, 6)

The ingredients were combined according to standard procedure. The coating was pigmented with titanium dioxide as the sole pigment. The list and order of addition is shown in Table 1.

Polyurethane coating (04, 4)

These formulations were prepared as follows. Bayhydrol XP7093 a polyester emulsion resin, antimicrobial agent A and the pigment were mixed together until a homogenous mixture was obtained. The pigment was ground and dispersed for 45 minutes. A solution of 1.0 g BYK 333 in 6.0 g of butanol was made and added to the resin-pigment mixture. The Bayhydur 302 was added slowly over a period of 10 minutes to the resin mixture while being mixed

Table 1. Representative pigmented latex formulation

Formulation	06	6	6_2
Material	Amounts (g)	Amounts (g)	Amounts (g)
Water	108	108	108
Tego 750 W	30.97	30.97	30.97
BYK 023	3.62	3.62	3.62
Ti Pure 960	400	400	200
Neocryl A-622	512	512	512
Igepal Co. 630	2.32	2.32	2.32
Ammonia 28%	1.6	1.6	1.6
Texnol	20	20	8.59
Polyphobe TR 116	1.98	1.98	1.98
Polyphobe TR 117	2.45	2.45	2.45
Antimicrobial agent A		21.66	
Antimicrobial agent B			21.66

with a mechanical stirrer. The viscosity was adjusted to between 20-25 seconds in a Zahn #2 cup. After adjusting the viscosity the formulation was filtered with a 50-micron filter and sprayed onto glass and steel panels, no sediment was collected. The panels were left to flash off for 15 minutes before baking at 80°C for 30 minutes. A typical formulation of the pigmented polyurethane coating is shown in Table 2.

Physical properties

Some physical properties of coatings prepared for antimicrobial testing are shown in Table 1. The air dry coatings were allowed to dry for 7 days at 25°C for 7 days before testing. High temperature cured coatings were also conditioned for at least 24 hours before testing. The solvent resistance was tested according to ASTM D 5402-93. The adhesion test was done according to ASTM D 3359-92a. The surface roughness testing was carried out by using a laboratory Profilometer. Key physical properties are listed in Table 3.

Accelerated Environmental Exposure

To simulate in service contamination, minute amounts of commercial food grade corn oil was diluted in xylene and the test coupons were wiped with this solution and allowed to dry at room temperature. The average oil loading level was measured to be approximately 10-20μg/ cm^2.

Table 2. Pigmented polyurethane coating formulation

Formulation	04	4
Material	Amounts (g)	Amounts (g)
Bayhydrol XP 7093	70.18	70.09
Antimicrobial agent A	-	1.08
Phthalo blue	5.08	5.18
BYK 333 in butanol	7.05	7.08
Bayhydur 302	31.0	31.0
Water	43	41

Table 3. Key Physical Properties of Prototype Antimicrobial Coatings

Run	Adhesion	Film thickness (μm)	Pencil Hardness	70% ETOH Resistance	Appearance
01	5B	45	3B	> 200 rubs	Clear
1_A	5B	50	3B	> 200 rubs	Hazy
1_B	5B	42	3B	> 200 rubs	Clear
02	5B	45	3B	> 200 rubs	Clear
2_A	5B	45	3B	> 200 rubs	Clear
2_B	5B	50	3B	> 200 rubs	Clear Pink*
06	4B	40	2B	> 200 rubs	White
6_A	4B	45	2B	>·200 rubs	White
6_B	4B	45	2B	> 200 rubs	White
6_B	3B	45	3B	> 200 rubs	White
04	5B	40	2H	> 200 rubs	Blue
4_A	5B	45	2H	> 200 rubs	Blue
05	5B	50	2H	> 200 rubs	Green
5_A	5B	50	2H	> 200 rubs	Green

01: control
$1_{A,B}$: Antimicrobial agents
* Changed color on standing

Similarly, test coupons were stored at room temperature in a draft-free office environment for 60 days before testing. Ultraviolet light exposure was done according to ASTM D4587 Cycle D. The cycle calls for the use of QUV-A340 bulbs exposure for 8 hours of UV light at 60°C and 4 hours of condensation at 45°C. The coatings were exposed to the UV light for 40 days.

Heat conditioning was carried out by storing the test coupons in a laboratory oven at 40 °C for 40 days. Humidity exposures were carried out by storing the test specimens in a controlled humidity cabinet at 70% relative humidity for 40 days.

Sterilization of coupons

The antibacterial coatings to be tested were supplied as a single-sided coating on standard 1" X 2" glass microscope slides. Sterilization of the coupons to rid them of any contaminating microorganisms prior to the challenge testing was achieved after exposure of the coupons to the germicidal UV lamp in a Nuaire Model NU-455-600 Class II Type A/B3 laminar flow hood (2 min. on each side at RT).

Preparation of Bacterial Broth Cultures

Escherichia coli ATCC 11229, *Staphylococcus aureus* ATCC 6538, *Bacillus subtilis* ATCC 6633, and *Pseudomonas aeruginosa* ATCC 15442 (all bacteria obtained from: Culti-loops, Remel Europe, Dartford, Kent, UK) were selected for the challenge testing of the antimicrobial coatings. Each bacterial strain was streaked for isolated colonies on Tryptic Soy Agar plates (TSA; Difco, Becton Dickinson, Sparks, MD) and incubated at 37°C for 12 to 18 hours. Broth cultures were prepared by first picking 1-2 isolated colonies from TSA plates and inoculating 5 ml Tryptic Soy Broth (TSB; Difco, Becton Dickinson, Sparks, MD) in a large test tube. This culture was grown in a shaking water bath for 8 hours at 37°C. After the eight hours, 125 µl of this culture is added to 50 ml of TSB in a side-arm Klett flask and grown for approximately 8 hours in shaking water bath at 37°C. The bacterial broths were grown to an A_{590} of 1.2-1.5 which correlated to approximately 10^9 CFU (colony forming units)/ ml. 25 ml of this broth was transferred to a sterile 30 ml centrifuge tube and centrifuged at 10,000 rpm for 10 minutes at 4°C (Sorval SS34 rotor) to pellet the bacteria. The bacterial pellet was washed one time and re-suspended in 25 ml of Phosphate Buffer Saline (PBS; BBL, FTA Hemagglutination Buffer, Becton Dickinson, Sparks, MD). The suspension was diluted with PBS to achieve the target range of 10^5-10^6 CFU/ml and used to inoculate the coupons containing antibacterial coatings.

Application of Bacteria to Coating

All inoculations were performed in a laminar flow hood. Sterilized coupons were placed into a sterile petri dish. 0.5 ml of prepared bacterial suspensions was pipetted directly onto the center of each coupon which formed a drop approximately 1.5 cm in diameter. A total of 36 coupons were tested in each individual run of the experiment; Table 4 details the experimental design for each coating tested.

Table 4. Design of a Typical Antibacterial Coating Challenge Test.

Inoculum	0 hour[1,2]	12 hour[1,2]	24 hour[1,2]
E. coli ATCC11229	2 w/AMA, 2 w/o AMA	2 w/AMA, 2 w/o AMA	2 w/AMA, 2 w/o AMA
S. aureus ATCC6538	2 w/AMA, 2 w/o AMA	2 w/AMA, 2 w/o AMA	2 w/AMA, 2 w/o AMA
B. subtilis ATCC 6633	2 w/AMA, 2 w/o AMA	2 w/AMA, 2 w/o AMA	2 w/AMA, 2 w/o AMA
P. aeruginosa ATCC 15442	2 w/AMA, 2 w/o AMA	2 w/AMA, 2 w/o AMA	2 w/AMA, 2 w/o AMA
PBS only	2 w/AMA, 2 w/o AMA	2 w/AMA, 2 w/o AMA	2 w/AMA, 2 w/o AMA

[1]w/AMA = coating with the antimicrobial agent
[2]w/o AMA = coating without antimicrobial agent

Once the coupons were inoculated, those assigned to incubate for 12 or 24 hours were placed in a humidified chamber. This was achieved by placing the petri dishes containing the coupons on test tubes that had been taped to the bottom of a 9"x13" Pyrex dish. These test tubes held the petri dishes just above the bottom of the dish, into which ~500 ml of sterile water was added. The dish was then covered with plastic wrap and tin foil and incubated at room temperature (20°C to 25°C).

Resuspension of Remaining Bacteria

At each time point, 25 ml of sterile PBS was pipetted into the petri dish completely submerging the coupon. A sterile plastic inoculating loop was used to release any viable bacteria remaining on the surface of the coating, and to mix the contents of the petri dish. Samples from this bacterial suspension were diluted in PBS, plated onto TSA plates, and incubated at 37°C for at least 24 hours. Plates containing between 30 and 300 isolated colonies were counted and used to calculate remaining bacterial concentrations.

Results and Discussion

Quantative Recovery of the Bacterial Inoculum from Coated Coupons

The bacterial strains and techniques used in this study were selected based on their use in testing disinfectants by various standard methods (36, 41, 37-40, 42). Coupons containing pigmented coatings with no antimicrobial agent (AMA) were inoculated with 0.5 ml of 10^5-10^6 CFU/ml and incubated according to the conditions outlined in the "Materials and Methods" section. Figure 1 demonstrates a nearly quantitative recovery of bacteria at all three time-points. These data suggested that the bacterial cells were not tightly bound to the coating's surface and that the bacteria could be easily removed by simple washing and abrasion of the slide in PBS. Detailed experiments, however, testing the adherence of the bacteria to the coating's surface have not been performed. Unless otherwise noted the selected test organisms are assumed to be *E. coli* ATCC 11229 and *S. aureus* ATCC 6538 and the coatings provided contained a standard 2.0% antimicrobial agent.

Time Course of Antimicrobial Killing

Studies by Galeano et al. and Matsumura et al. (42-43) suggested that bactericidal action could be seen as early as 2 hours after bacterial contact with their antimicrobial coating. In our studies, we performed a time course looking at bacterial killing at 1, 6 and 12 hours after the inoculation of the coated coupons. Figure 2 outlines the results produced at these early time points. There were no significant reductions observed before 12 hours.

Vegetable and Natural Oil Loading

Many surfaces over time will obtain a layer of dust and other substances that may interfere with the efficacy of the coating. Two experiments were set up to test the effects of organic loads on the coating. First, one set of coupons was coated with a layer of vegetable oil and tested. A second set was exposed to the air in a lab to create a natural load on the coating. Another set of coupons containing the same number of coupons left in the lab environment was separately sealed in a dessicator during the time of exposure to act as the controls for this experiment. The natural load produced by the air in the lab did not seem to affect the efficacy of the coating when compared to those sealed in the dessicator. They maintained 100% killing at 12 hours for both *E. coli* and *S. aureus*. For the coupons with the vegetable oil load, *E. coli* was completely killed by 12 hours in all three runs. *S. aureus* had varying results though. The

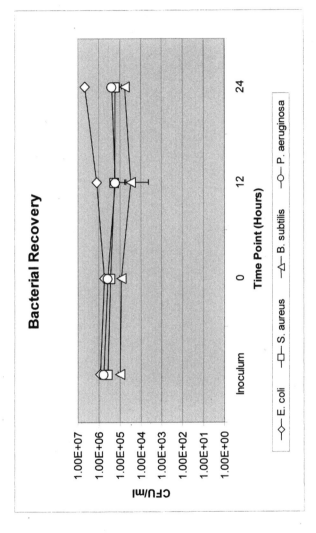

Figure 1. CFU/ml of surviving bacteria on coatings lacking any antimicrobial agent (n = 3 experiments for each organism).

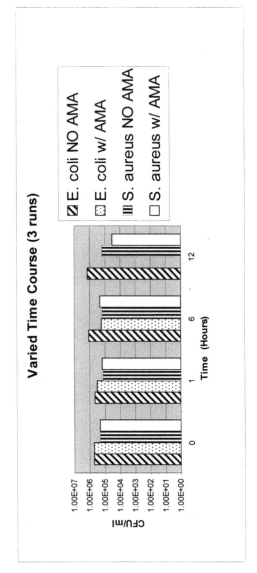

Figure 2. CFU/ml of surviving bacteria at early time points

first run showed 100% complete killing after 12 hours, the second showed no reduction at 12 hours and a 3 log reduction at 24 hours, and the third run produced a 1 log reduction after 12 hours and 100% killing after 24 hours. These data suggest that the inert vegetable oil was in some way interfering with the coatings ability to kill this gram positive organism.

Challenge of coating using *Bacillus subtilis* and *Pseudomonas aeruginosa*

Once we were confident with the results using *E. coli* and *S. aureus*, two other organisms were introduced to test the range of effectiveness of the coating. The two newly selected organisms for these runs were the gram positive, *B. subtilis* ATCC6633, and the gram negative, *P. aeruginosa* ATCC15442. The coating showed the ability to completely kill both of these bacteria after 12 hours of incubation which was consistent with the results observed with the first two organisms.

Ceramic substrate and repeatability

To determine if the substrate that the coating was attached to affected the antibacterial properties a set of ceramic tiles were completely covered with the coating on all sides and then run through the challenge. Upon completing this first run it was observed that the adherence of the coating to the ceramic was much improved than when comparing to the glass slides used in earlier tests. This allowed the tiles to be recovered, dried, re-sterilized and tested again to answer the question of repeatability. The ceramic coupons were tested a total of three times. The first run showed 100% killing of *E. coli* after 12 hours. The 2nd and 3rd runs produced reductions by 12 hours and 100% killing by 24 hours for *E. coli*. Each of the three runs showed reductions of *S. aureus* after 12 hours and 100% killing by 24 hours. The reductions observed during the repeat tests using the two organisms after 12 hours averaged only a 1-2 logs, showing a slight decrease in efficacy from the initial challenge of these coatings.

Varied Conditions

To determine the effects of conditions a particular coating may endure after application a number of different tests were performed on the coating prior to the antibacterial challenge. These included exposing the coupons to UV radiation, increased levels of humidity, temperature, or application of a common cleaning agent (Windex). The challenge was then conducted to determine if there had been any change in antibacterial efficacy.

The complete reduction of *E.coli* after 12 hours was consistent on all treated coatings when compared to prior tests on standard coatings provided at normal conditions. The expected reduction of *S. aureus* at 12 hour and complete reduction at 24 hour was observed for coupons exposed to increased humidity or temperature and those cleaned with the Windex. The UV treated coatings showed a decrease in effectiveness against *S. aureus* though. There was no reduction observed after 12 hours and only a 2 log reduction after 24 hours.

Dose Response with Increasing Levels of Bacteria

These experiments were performed to determine the efficacy of bacterial killing with increasing levels of bacterial load. Using the standard bacterial load of 10^5 CFU/ml, 100% killing was observed at both 12 hour and 24 hour (Table 5). Increasing the bacterial inoculum to 10^7 CFU/ml demonstrated significant killing of *S. aureus* (>87%) at 12 hour and complete killing at 24 hour. The *E. coli* was completely killed at 12 hour and 24 hour using this same inoculum of 10^7 (Table 5). Using a 10^9 CFU/ml inoculum, however, required 24 hours to achieve 31% and 72% killing of *E. coli* and *S. aureus*, respectively (Table 5). The results when the time course for this concentration of inoculum is extended to 48 and 72 hours of incubation can be seen in Table 5.

Conclusions

Silver-based antimicrobial coatings have been shown to be effective in killing a variety of bacteria. In the present studies, we have demonstrated that air-dry waterborne coatings containing silver-based antimicrobials effectively kill the test strains of *B. subtilis* and *P. aeruginosa* after 12 hours of incubation at RT. In experiments employing a varied time course it was demonstrated that no significant reductions were occurring before 12 hours. When the coating was allowed to collect a "natural organic load" produced by exposure to a lab environment, there was no difference in bacterial killing. When vegetable oil was used as an organic load there was no effect on the reduction of *E. coli*, but the killing of *S. aureus* was inhibited. Application of the coating to a different substrate, ceramic tiles, did not seem to change the antibacterial efficacy of the coating. When the challenge was repeated on the same coating for a total of three runs, the results showed a slight decrease in effectiveness after 12 hours of incubation when compared to the initial run. We have also showed that the coating remains effective when exposed to increased temperature, humidity and a common cleaning agent, Windex. When exposed to UV radiation however, there was an observed decrease in efficacy against *S. aureus*. Increasing the bacterial inoculums from 10^5 to 10^7 showed no difference in killing after 24 hours. When the concentration of bacteria was increased to 10^9 there was a significant difference in the coatings ability to kill the bacteria. After 24 hours

Table 5. Dose Response using Increasing Levels of Bacteria

Time (Hours) CFU/ml	0		12		24		48		72	
	E. coli	S. aureus	E. coli	S. aureus	E. coli	S. aureus	E. coli	S. aureus	E. coli	S. aureus
~10^5	0	0	100	100	100	100	n/a	n/a	n/a	n/a
~10^7	0	0	100	87	100	100	n/a	n/a	n/a	n/a
~10^9	0	0	31	36	31	72	80	100	81	99

All data reported as percentage killed
[1] E. coli ATCC11229 and S. aureus ATCC 6538
[2] % killing was determined using the remaining CFU/ml at each time point and dividing by the CFU/ml at T_0.

at this bacterial concentration only 31% of *E. coli* and 72% of *S. aureus* were killed. The proposed possible applications of such coatings include areas such as clinical settings, food preparation areas, and food processing facilities.

References

1. US Pat. 6,228491, Fibrous textile articles possessing enhanced antimicrobial properties.
2. Herrera, M, Carrion, p. et. al. In vitro antibacterial activity of glass-ionomer cements, Microbios, 104, 409, 2001.
3. Cho, D. L, et. al., A study on the preparation of antimicrobial biopolymer film, J.Microbiol. Biotechnol. 11(2), 193-198, 2001.
4. US. Pat Appl. 2000-57-2716, Milliken company, USA. Antimicrobial transfer substrate for textile finishing.
5. Alexander, M. K., Klibanov, J., Lewis, C., Teller, Chung-Jen Liao, Proc. Natl.Acad. Sci., 98, 5981, 2001.
6. Alexander, M. K. Klibanov, J. Lewis, C. Teller, Chung-Jen Liao, S.B. Lee, Biotechnol. Lett., 24, 801, 2002.
7. Alexander, M., K., Klibanov, J. Lewis, C. Teller, Chung-Jen Liao, S.B. Lee, Biotechol. Bioeng., 79, 466, 2002
8. Klein; M., W. F. DeGardo; Proc. Natl. Acad. Sci. 99, 5110, 2002,
9. Mackeen, C, S Person, SC Warner, W Snipes and SE Stevens, Jr. Silver-coated nylon fiber as an antibacterial agent. Antimicrob. Agents Chemother. 31, 93-99, 1987.
10. Schierholz, J.M., Z. Wachol-Drewek, L.J. Lucas and G. Pulverer. Activity of silver ions in different media. Zent.bl. Bakteriol. 287, 411-420, 1998.
11. Spadaro, A. Berger, S.D. Barranco, S.E. Chapin and R.O. Becker. Antibacterial effects of silver electrodes with weak direct current. Antimicrob. Agents Chemother. 6, 637-642, 1974.
12. Thibodeau, E.A., S.L. Handelman and R.E. Marquis. Inhibition and killing of oral bacteria by silver ions generated with low density electric current. J. Dent. Res. 57, 922-926, 1978.
13. Klasen, H.J. Historical review of the use of silver in the treatment of burns. I. Early uses. Burns 26:117-130, 2000.
14. Klasen, H.J., A historical review of the use of silver in treatment of burns, II. Renewed interest for silver, *Burns*, 26, 131-138, 2000.
15. Maki, D.G., and P.A. Tambyah, Engineering out the risk of infection with urinary catheters. Emerg. Infect. Dis. 7:1-6, 2001.
16. Bromberg, L.E., V.M. Braman, D.M. Rothstein, P. Spacciapoli, S.M. O'Connor, E.J. Nelson, D.K. Buxton, M.S. Tonetti, and P.M. Friden. Sustained release of silver from peridontal wafers for treatment of periodontitis. J. Controlled Rel. 68:63-72, 2000.

17. Quintavalla, S., and L. Vicini. Antimicrobial food packaging in meat industry. Meat Science 62:373-380, 2002.
18. Wright, J.B., K. Lam, D. Hansen, and R.E. Burrell. Efficacy of topical silver against fungal burn wound pathogens. Am. J. Infect. Control 27:344-350, 1999.
19. von Naegelli, V. Deut. Schr. Schweiz. Naturforsch. Ges. 33, 174-182. 1893.
20. Clement, J.L. and P.S. Jarrett. Antibacterial silver. Metal-Based Drugs 1, 467-482, 1994.
21. Thurman, R.B. and C.P. Gerba. The Molecular Mechanisms of Copper and Silver Ion Disinfection of Bacteria and Viruses. CRC Critical Reviews in Environmental Control 8, 295-315, 1989.
22. Yahya, M.T., L.K. Landeen, M.C. Messina, S.M. Kute, R. Schulz, and C.P. Gerba. Disinfection of bacteria in water systems by using electrolytically generated copper:silver and reduced levels of free chlorine. Can. J. Microbiol. 36,109-116,1990.
23. Becker, R.O. Silver ions in the treatment of local infections. Metal-Based Drugs 6, 311-314, 1999.
24. Carr, H.S., T.J. Wlodkowski and H.S. Rosenkranz. Silver sulfadiazine: In vitro antibacterial activity. Antimicrob. Agents Chemother. 4, 585-587, 1973.
25. Yoshida, K., M. Tanagawa, M. Atsuta. Characterization and inhibitory effect of antibacterial dental resin composites incorporating silver-supported materials. J. Biomed. Mater. Res. 47, 516-522, 1999.
26. Dibrov, P., J. Dzioba, K.K. Gosink, and C.C. Hase. 2002. Chemiosmotic mechanism of antimicrobial activity of Ag+ in Vibrio cholerae. Antimicrob. Agents Chemother. 46:2668-2670.
27. Gupta, A., L.T. Phung, D.E. Taylor, and S. Silver. 2001. Diversity of silver resistance genes in IncH incompatibility group plasmids. Microbiol. 147:3393-3402.
28. Franke, S., G. Grass, and D.H. Hies. the product of the ybdE gene of the Escherichia coli chromosome is involved in detoxification of silver ions. Microbiol. 147:965-972.
29. Nies, D.H. 1999. Microbial heavy metal resistance. Appl. Microbiol. Biotechnol. 51:730-750.
30. Gupta, A., M. Maynes, and S. Silver. 1998. Effects of halides on plasmid-mediated silver resistance in Escherichia coli. Appl. Environ. Microbiol. 64:5042-5045.
31. Silver, S., A. Gupta, K. Matsui, and J.-F. Lo. 1999. Resistance to Ag(I) cations in bacteria: environments, gene, and proteins. Metal-Based Drugs 6:315-320.
32. Slawson, R.M., E.M. Lohmeier-Vogel, H. Lee, and J.T. Trevors. 1994. Silver resistance in Pseudomonas stutzeri. BioMetals 7:30-40.
33. Deshpande, L.M., and B.A. Chopade. 1994. Plasmid mediated silver resistance in Acinetobacter baumannii. BioMetals 7:49-56.

34. Hendry, A.T., and I.O. Stewart. 1979. Silver-resistant enterobacteriaceae from hospital patients. Can. J. Microbiol. 25:915-921.
35. Barkley, W.E., and J.H. Richards. Laboratory Safety. Laboratory safety, p. 715-734 *In* P. Gerhardt, R.G.E. Murray, W.A. Wood, and N.R. Krieg (ed.), *Methods for General and Molecular Bacteriology*, American Society for Microbiology, Washington, D.C. 1994.
36. Luppens, S.B.I., M.W. Reij, R.W.L. van der Heijden, F.M. Rombouts, and T. Abee. Development of a standard test to assess the resistance of *Staphylococcus aureus* biofilm cells to disinfectants. Appl. Environ. Microbiol. 68, 4194-4200. 2002.
37. Rubino, J.R., J.M. Bauer, P.H. Clarke, B.B. Woodward, F.C. Porter, and H.G. Hilton. Hard surface carrier test for efficacy testing of disinfectants-collaborative study. J. AOAC Int. 75, 635-645. 1992.
38. Mariscal, A., M. Carnero-Varo, J. Gomez-Aracena, and J. fernandez-Crehuet. Development and testing of a microbiological assay to detect residual effects of disinfectant on hard surfaces. Appl. Environ. Microbiol. 65, 3717-3720. 1999.
39. Abrishami, S.H., B.D. Tall, T.J. Bruursema, P.S. Epstein, and D.B. Shah. Bacterial adherence and viability on cutting board surfaces. J. Food Safety 14, 153-172, 1994.
40. Miner, N., M. Armstrong, C.D. Carr, B. Maida, and L. Schlotfeld. Modified quantitative association of official analytical chemists sporicidal test for liquid chemical germicides. Appl. Environ. Microbiol. 63, 3304-3307, 1997.
41. Official Methods of Analysis. 15th Ed., Association of Official Analytical Chemists, Arlington, VA., Method 955.15, 961.02, 964.02, 965.19, and 966.04. 1990.
42. Galeano, B., E. Korff, and W.L. Nicholson. 2003. Inactivation of vegetative cells, but not spores, of *Bacillus anthracis*. *B. cereus*, and *B. subtilis* on stainless steel surfaces coated with an antimicrobial silver- and zinc-containing zeolite formulation. Appl. Environ. Microbiol. 69:4329-4331.
43. Matsumura, Y., K. Yoshikata, S. Kunisaki, and T. Tsuchido. 2003. Mode of bactericidal action of silver zeolite and its comparison with that of silver nitrate. Appl. Environ. Microbiol. 69:4278-4281.

Chapter 6

Novel, Environmentally Friendly, Antifouling/Fouling Release Coatings Developed Using Combinatorial Methods

Bret J. Chisholm, David A. Christianson, Shane J. Stafslien, Christy Gallagher-Lein, and Justin Daniels

Center for Nanoscale Science and Engineering, North Dakota State University, Fargo, ND 58102

Novel, environmentally-friendly, antifouling coatings were prepared by incorporating either triclosan or ammonium salt functionality into moisture-curable polysiloxane-based coatings. The biocidal functionality was covalently bound to the coating matrix to prevent leaching of toxic biocide to the environment. High throughput biological assays based on marine bacteria growth and settlement were used to characterize the antifouling character of the coatings and prove that antifouling character was not due to a leaching effect. The minimum amount of tethered biocide required to deter settlement of marine bacteria was determined for two different classes of moisture-curable coatings.

Introduction

Fouling of ship hulls by marine organisms is a serious problem in terms of fuel efficiency, vessel maneuverability and speed, engine lifetime, and marine ecology. A study by the US Navy showed that light fouling on a vessel can cause a 15 percent increase in fuel consumption while heavy fouling may cause as much as a 45 percent increase in fuel consumption. (*1*) With regard to marine ecology, transport of organisms from one harbor to another by a fouled ship hull can cause changes in marine ecosystems around the world.

To date, the most effective method of preventing or reducing marine fouling is to incorporate leachable biocidal compounds into hull coatings. Very effective antifouling coatings have been created by incorporating organotin compounds into polyacrylate coatings by copolymerizing tributyltin methacrylate with methyl methacrylate. (*2*) The coatings deter settlement of marine organisms by the slow release of tributyltin via hydrolysis of tin ester groups. While coatings based on organotin compounds have been shown to be very effective, they have been banned from use due to the adverse effect they have on the environment. (*3*) Currently, the most prevalent coatings on Naval vessels are comprised of a polyacrylate matrix filled with copper oxide as a biocide. (*4*) In addition to copper oxide, organic booster biocides are often used. While less toxic to the environment than organotin-based coatings, copper-containing coatings are also an environmental hazard. (*5,6*) As a result, an extensive effort has been ongoing to develop new environmentally-friendly coatings for ship hulls.

One environmentally-friendly approach to combating marine fouling has been to create coating surfaces that allow for easy release of fouling organisms by minimizing the adhesive strength between marine organisms and the coating surface. (*7*) This has been achieved by creating low surface energy, elastomeric surfaces using polysiloxane coatings. (*8*) Easy removal of marine organisms has been demonstrated for some polysiloxane-based coatings. However, the poor durability of these relatively soft coatings has limited commercial acceptance of this class of foul-release coatings.

The approach being investigated in our laboratory for generating new environmentally-friendly coatings for ship hulls involves combining characteristics of both antifouling and foul-release coatings by tethering biocidal moieties to a low surface energy, elastomeric coating matrix. (*9*) The organic biocidal moieties deter settlement of most marine organisms while the low surface energy, elastomeric matrix allows for easy release of those organisms capable of settlement. The concept of tethering the biocide to the coating matrix through stable, covalent bonds alleviates issues associated with leaching of toxic components into the marine environment.

To facilitate the material development effort and reduce the time associated with material discovery and optimization, a combinatorial development approach was adopted. (*10*) A combinatorial approach involves aspects of automation, miniaturization, and database management to increase the rate of experimentation at least an order of magnitude. In our laboratory, we have developed a combinatorial workflow for marine coating development that involves all aspects of the experimental process, as shown in Figure 1. The details of this workflow have been provided elsewhere. (*11*)

Experimental

Materials

Materials used for the preparation of coatings are described in Table I. All materials were used as received from the manufacturer. Allyl triclosan was prepared as described by Thomas et al. (*9*)

Dimethylsiloxane-methylhydrosiloxane copolymer (PDM-co-HS) was functionalized with triclosan and methoxysilane groups by the sequential hydrosilylation of allyl triclosan and allyltrimethoxysilane with PDM-co-HS. Allyltriclosan and PDM-co-HS (see Table II) were dissolved in 10 ml. toluene and 40 mg. platinum(0)-1,3-divinyl-1,1,3,-tetramethyldisiloxane complex (Karstedt's catalyst) added to the solution. The solution was refluxed for 24 hours under a nitrogen blanket. After cooling to room temperature, allyltrimethoxysilane (Table II) was added and the reaction mixture allowed to stir at 40 °C for 24 hours. Coating solutions were prepared by adding 1.0 weight percent tetrabutylammonium fluoride to the polymer solutions as a condensation catalyst.

Table I. Description of materials used to prepare coatings

Chemical Name or Acronym	Supplier	Grade
Allyltrimethoxysilane	Gelest	----
Dimethylsiloxane-methylhydrosiloxane copolymer (PDM-co-HS)	Gelest	HMS-301
Platinum(0)-1,3-divinyl-1,1,3,-tetramethyldisiloxane complex	Aldrich	----
Toluene	VWR	----
Trimethoxysilylpropyl modified polyethylenimine (PEI-H$^+$Cl$^-$)	Gelest	SSP-060
Silanol-terminated polydimethylsiloxane (HOSi-PDMS-SiOH)	Gelest	DMS-S35
Tetrabutylammonium fluoride	Aldrich	----
4-methyl-2-pentanone	Aldrich	----

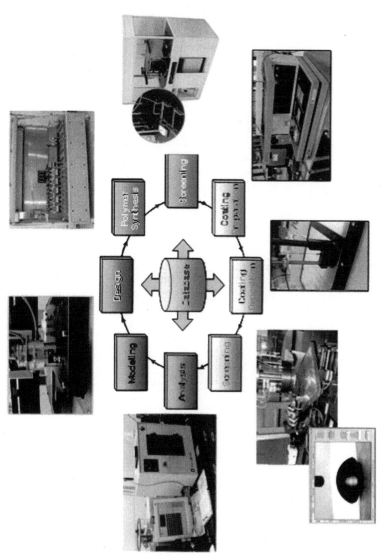

Figure 1. An illustration of the combinatorial workflow for polymer and coating development at North Dakota State University. (See page 1 of color insert.)

Table II. Reagent concentrations used to prepare triclosan and methoxysilane functional polysiloxanes. All values are in grams

Sample ID	wt. HMS-301	wt. allyltriclosan	wt. allyltrimethoxysilane
0 wt. % triclosan	2.30	0	0.38
10 wt. % triclosan	2.10	0.28	0.35
22 wt. % triclosan	1.90	0.64	0.31
30 wt. % triclosan	1.70	0.86	0.28
37 wt. % triclosan	1.50	1.01	0.25
46 wt. % triclosan	1.30	1.31	0.22

Moisture-curable coatings derived from silanol-terminated polydimethyl-siloxane (HOSi-PDMS-SiOH) and acidified trimethoxysilylpropyl modified polyethylenimine (PEI-H$^+$Cl$^-$) were prepared by solution blending in 4-methyl-2-pentanone and using 1.0 weight percent tetrabutylammonium fluoride as a condensation catalyst.

Bacterial Growth and Retention Assay

Each experimental coating was cast into a row of a modified 24 well plate resulting in 6 replicate samples. Upon curing, plates were immersed in a circulating deionized water bath for 7 days of preconditioning to facilitate removal of any toxic components that may be present and potentially interfere with bacterial analysis. Preconditioned coatings were then incubated in 1.0 mL of a nutrient limited growth medium (0.5 g peptone and 0.1 g yeast extract / 1L of artificial sea water) overnight (~18 hr) and the resultant coating leachate was collected. Each coating extract was then analyzed for toxicity via the addition of the marine bacterium *Cytophaga lytica*[12] (50 µl of ~10^{7}cells/mL / 1.0 mL leachate) and overnight incubation at 28°C. Turbidity measurements were then made (600nm) to assess the amount of biofilm growth obtained in each coating leachate. A significant reduction in the amount of biofilm growth was considered to be a consequence of toxic components leaching into the medium solution.

The antifouling performance of each coating formulation was evaluated by assessing its' ability to inhibit or minimize bacterial biofilm growth and retention. The formulations cast into the 24 well plates were inoculated with stationary phase cultures of the marine bacteria *Cytophaga lytica*, *Halomonas pacifica* ATCC 27122, *Cobetia marina* ATCC 25374, and *Pseudoalteromonas atlantica* ATCC 19262 (~24 hr) resuspended in a nutrient limited growth medium at ~10^{7}cells/mL. The plates were then placed in a 28°C incubator for 18-48 hours to facilitate bacterial attachment and subsequent colonization. The plates were then rinsed three times with 1.0 mL of deionized water to remove any planktonic or loosely attached bacterial biomass. The amount of biofilm retained on the coating surfaces was then stained with the biomass indicator dye crystal violet (0.3%). Digital images were taken of each plate and coating formulations were analyzed for percent surface coverage via cropping and grayscale conversion in AdobePhotoshop (Adobe Systems Inc.) and analysis in PhotoGrid 1.0 (University of Hawaii). The percent of biofilm coverage on each surface was determined by evaluating the ratio of dark to light pixels. Once dry, the crystal violet dye was extracted from the biofilm with addition of 0.5 mL of glacial acetic acid (33 wt%) and the resulting eluate measured for absorbance at 600 nm. The measured absorbance values obtained were directly proportional to the amount of biofilm retained on the coating surface.

Results and Discussion

A number of organic compounds have been shown to exhibit biocidal activity towards an array of marine organisms. (*13*) The mechanism of interaction between the biocide and the organism varies as a function of the biocide chemical structure. For our studies, we chose to investigate the use of 5-chloro-2-(2,4-dichlorophenoxy)phenol (triclosan) which is a widely-used, relatively mild, EPA-registered biocide. (*14*)

We have previously shown that covalently linking, i.e. tethering, triclosan to a polydimethylsiloxane (PDMS) polymer backbone deters settlement of marine organisms in ocean testing. (*9*) The coatings that were previously investigated were cured using an epoxy-amine cure in which epoxy groups are attached to the PDMS backbone.

More recently we investigated the incorporation of triclosan into moisture-curable siloxane coatings using the synthetic process shown in Figure 2. Initial experiments were designed to evaluate the effect of triclosan level on antifouling properties. Antifouling properties were determined using high throughput biological assays that have been described in detail elsewhere. (*15*)

The workflow for the biological assays is shown schematically in Figure 3. The leachate toxicity assay serves to ensure that any observed antifouling character is due to a coating surface associated phenomenon and not to the leaching of toxic components from the coating into the aqueous medium.

Figures 4 displays images of the array plates after conducting the antifouling assay. Each coating was replicated five times such that each row within an array plate represents a single coating composition. Since the bacterial biofilm produces a dark color due to the biofilm staining process, the reduction in biofilm coverage as a function of triclosan level in the coatings can be easily seen. Figure 5 displays a graphical representation of the percent coverage data as obtained from image analysis.

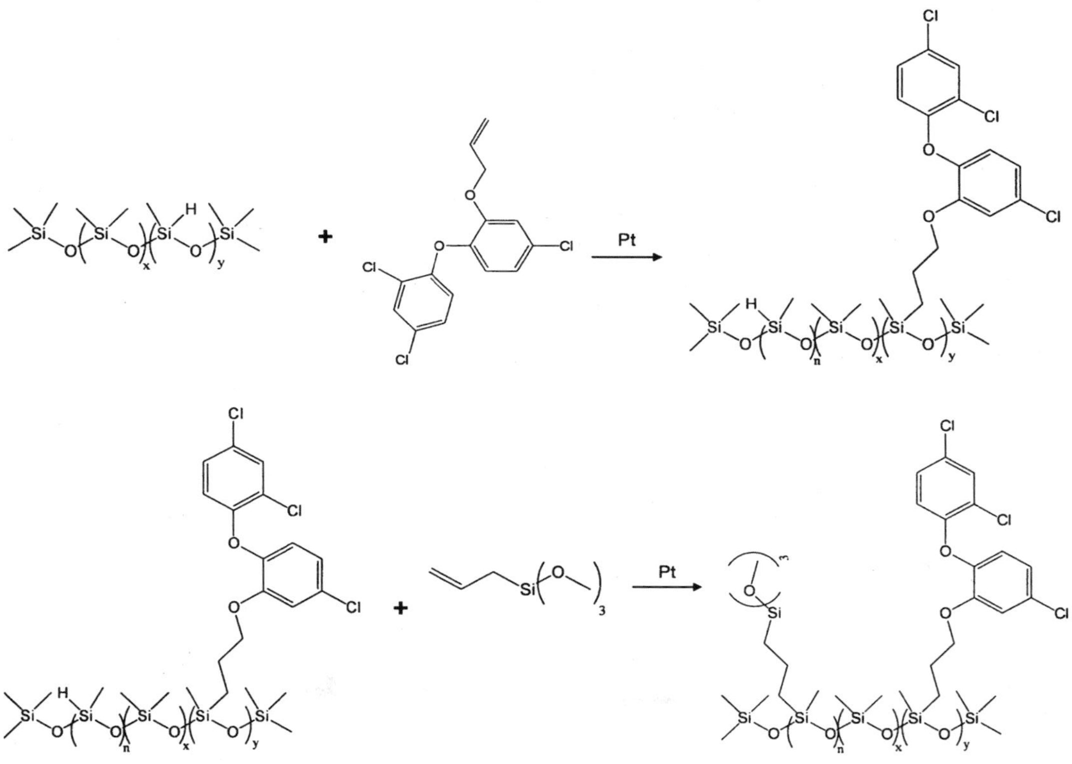

Figure 2. Schematic of the synthetic process used to prepare moisture-curable PDMS coatings containing tethered triclosan.

134

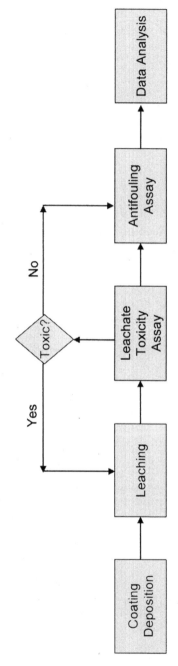

Figure 3. Schematic of the high throughput biological assays used to evaluate antifouling characteristics.

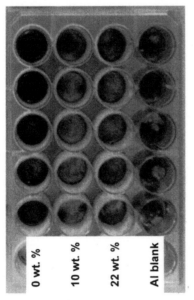

Figure 4. Images of the array plates after conducting the antifouling assay. Weight percentages indicate the weight percent of tethered triclosan in the coatings. (See page 1 of color insert.)

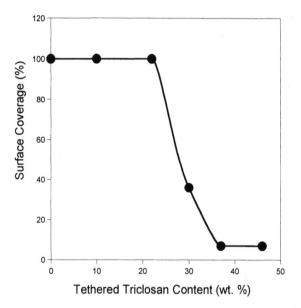

Figure 5. A graphical representation of the percent coverage data as obtained from image analysis. Weight percentages indicate the weight percent of tethered triclosan in the coatings.

In addition to percent coverage, the amount of biofilm present on the coating surface was measured by extracting the stained biofilm and measuring the optical density of the extract. The results of the biofilm mass measurement are shown in Figure 6. A comparison of the data obtained from the biofilm coverage measurement (Figure 5) and the biofilm mass measurement (Figure 6) indicates a biofilm redistribution phenomenon.(*16*)

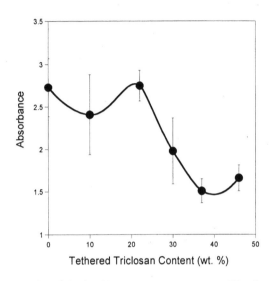

Figure 6. Results of the biofilm mass measurement. Weight percentages indicate the weight percent of tethered triclosan in the coatings.

Figure 7 displays the results of the leachate toxicity assay which clearly indicates that the antifouling results obtained for the coating containing 46 weight percent tethered triclosan were at least partially due to leaching of toxic components from the coating into the aqueous growth medium. This result was consistent with H^1 NMR results that showed a trace amount of unreacted allyl triclosan in the coating. Considering results from both the leachate toxicity assay and the bacterial coverage and mass assay, it can be concluded that the tethered triclosan deters settlement of *C. lytica* when present a concentration of at least 30 weight percent.

Another tethered biocide approach that is being investigated in our laboratory involves the use of bound ammonium salts in moisture-curable, siloxane-containing coatings. Ammonium compounds have been shown to be effective biocides.(*17*) Initial investigations within this class of materials utilized commercially available polymers. Figure 8 displays a schematic of the coating compositions explored. The first screening experiment involved an investigation of the effect of the level of the ammonium functional polymer

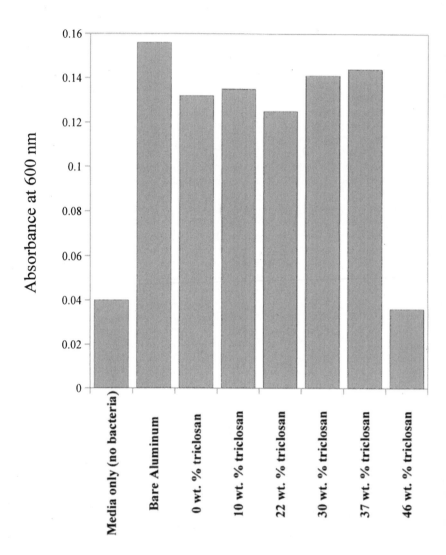

Figure 7. Leachate toxicity results.

(PEI-H⁺Cl⁻) on antifouling character using the bacterial biofilm growth assay. PEI-H⁺Cl⁻ levels from 0 to 25 weight percent of the total coating film composition were investigated. The results showed that the use of 10 weight percent or more PEI-H⁺Cl⁻ deterred biofilm growth on the coating surfaces. Leachate toxicity testing indicated that only the leachate obtained from the coating with the highest level of PEI-H⁺Cl⁻ possessed any toxicity. This result suggests that at this relatively high level of PEI-H⁺Cl⁻ some of the PEI-H⁺Cl⁻ polymer chains were not incorporated into the polymer matrix. Follow-up experiments utilizing a small molecule crosslinker to enhance crosslink density are planned to reduce leachate toxicity of coatings containing relatively high levels of PEI-H⁺Cl⁻. Figure 9 displays the reduction of bacterial biofilm growth obtained using 12.5 weight percent PEI-H⁺Cl⁻. At this level of PEI-H⁺Cl⁻, no leachate toxicity was observed and the dramatic reduction in bacterial biofilm growth can be attributed to antifouling character of the coating surface. It is interesting to note that this coating was effective toward all of the marine bacterial species utilized for the investigation as well as a bacteria of importance to humans, namely, *E. coli*.

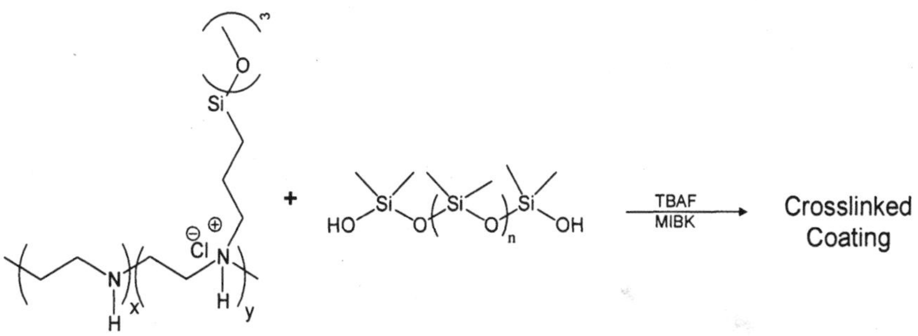

Figure 8. Synthetic scheme for moisture-curable coatings containing bound ammonium salt groups.

Conclusion

Tethering triclosan into moisture-curable polysiloxane coatings was found to deter settlement of marine bacteria when the level of triclosan was at least 30 weight percent of the total composition. For moisture-curable compositions based on PEI-H⁺Cl⁻, it was found that 10 weight percent or more PEI-H⁺Cl⁻ in the compositions deterred settlement of all of the marine bacteria evaluated. In addition to deterring the settlement of marine bacteria, these coatings also deterred the settlement of *E. coli*. Promising coating compositions from these studies are being scaled-up and will be evaluated using ocean site testing.

140

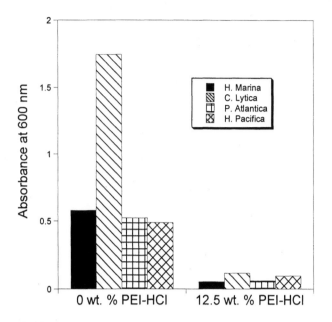

Figure 9. The effect of 12.5 weight percent PEI-H⁺Cl⁻ on biofilm growth using four different marine bacteria.

Acknowledgements

The authors thank the Office of Naval Research for financial support (grant N00014-04-1-0597).

References

1. Hull Cleaning. http://www.a1comdive.com/HullClean.htm
2. Yebra, D. M.; Kiil, S.; Dam-Johansen, K. *Prog. Org. Coat.* **2004**, *50*, 75-104.
3. Champ, M. A. *Mar. Pollut. Bull.* **2003**, *46*, 935-940.
4. Ferry, J. D.; Carritt, D. E. *Ind. Eng. Chem.* **1946**, *38*, 612-617.
5. Katranitsas, A.; Castritsi-Catharios, J.; Persoone, G. *Pollut. Bull.* **2003**, *46*, 1491-1494.
6. Terlizzi, A.; Fraschetti, S.; Gianguzza, P.; Faimali, M.; Boero, F. *Aquat. Conserv. Mar. Freshw. Ecosyst.* **1998**, *11*, 311-317.
7. Clare, A. S.; *Marine Biotech, J.* **1998**, *6*, 3-6.
8. Edwards, D. P.; Nevell, T. G.; Plunkett, B. A.; Ochiltree, B. C. *Int. Biodeterior. Biodegrad.* **1994**, *34*, 349-359.

9. Thomas, J.; Choi, S.-B.; Fjeldheim, R.; Boudjouk, P. *Biofouling.* **2004,** *20,* 227-236.
10. Chisholm, B. J.; Potyrailo, R.; Cawse, J.; Shaffer, R.; Brennan, M.; Molaison, C.; Whisenhunt, D.; Flanagan, B.; Olson, D.; Akhave, J.; Saunders, D.; Mehrabi, A.; Licon, M. *Prog. Org. Coat.* **2002,** *45,* 313-321.
11. Webster, D. C.; Bennet, J.; Kuebler, S.; Kossuth, M. B.; Jonasdottir, S. *JCT Coat. Tec.* **2004,** *1,* 34-39.
12. Huang, S.; Hadfield, M. G. *Mar. Ecol. Prog. Ser.* **2003,** *260,* 161-172.
13. Lewis, J. A. *Surf. Coat. Aust.* **2003,** *40,* 12-15.
14. Russell, A. D. *J. Antimicrobial Chemotherapy.* **2004,** *53,* 693-695.
15. Stafslien, S. J.; Bahr, J. A.; Feser, J. M.; Weisz, J. C.; Chisholm, B. J.; Ready, T. E.; Boudjouk, P. *J. Comb. Chem.* **2006,** *8,* 156-162.
16. Stafslien, S. J.; Bahr, J. A.; Feser, J. M.; Weisz, J. C.; Chisholm, B. J. *Biofouling,* unpublished.
17. Schaeufele, P. J.; *J. Am. Oil Chem. Soc.* **1984,** *61,* 387-389.

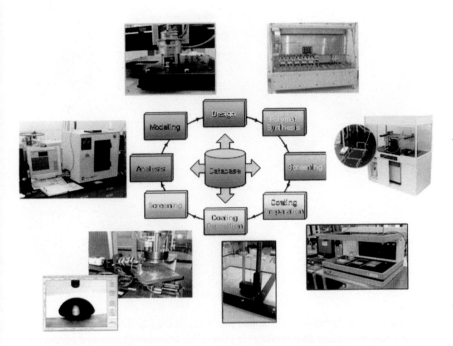

Figure 6.1. An illustration of the combinatorial workflow for polymer and coating development at North Dakota State University.

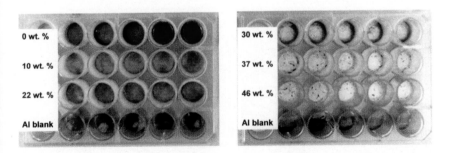

Figure 6.4. Images of the array plates after conducting the antifouling assay. Weight percentages indicate the weight percent of tethered triclosan in the coatings.

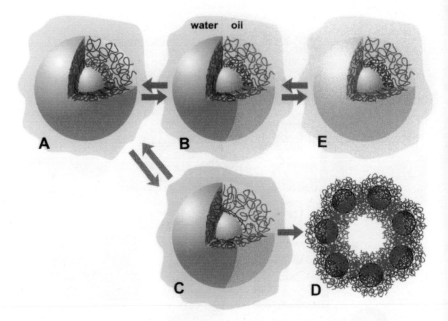

Figure 7.1. Schematic of the smart particles (mixed spherical brush grafted onto nanoparticles) that undergoes reversible transitions between core-shell morphologies (A, E) and Janus morphology at "water-oil" interface (B) and in nonselective solvent (C). The unsymmetrical morphology results in formation of supermicelles (D). Two unlike polymers are designated as red and blue chains.

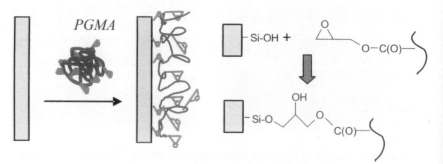

Figure 7.2. Reactive polymer attached to substrate.

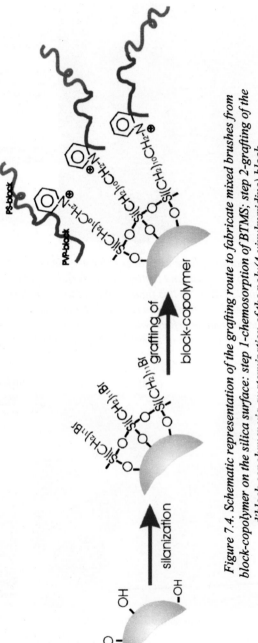

Figure 7.4. Schematic representation of the grafting route to fabricate mixed brushes from block-copolymer on the silica surface: step 1-chemosorption of BTMS; step 2-grafting of the diblock copolymer via quaternization of the poly(4-vinylpyridine) block.

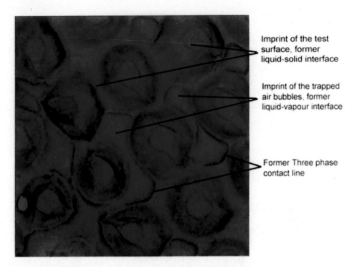

Imprint of the test
surface, former
liquid-solid interface

Imprint of the trapped
air bubbles, former
liquid-vapour interface

Former Three phase
contact line

Figure 13.8. The former phase interfaces ("former" here refers to the condition in when the drop was fluid and resting on the surface) can be identified on a gelled and removed drop. In this figure the contour of the former three-phase contact line is clearly visible on the superposed images, as is the shape of the former liquid–vapour interface with its nodes and loops. The large round features are imprints of spikes like those depicted in Figure 7.

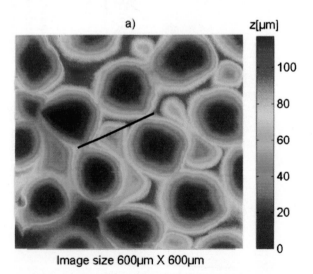

a)   z[μm]

100

80

60

40

20

0

Image size 600μm X 600μm

Figure 13.9a. This figure presents a height map of the underside of a drop taken along the black transect indicated in a). Using this method it is possible to obtain quantitative information concerning the geometry of the liquid–vapour interface.

Nanotechnology

Chapter 7

Coatings via Self-Assembly of Smart Nanoparticles

Sergiy Minko[1], Igor Luzinov[2], Mikhail Motornov[1], Roman Sheparovych[1], Robert Lupitskyy[1], Yong Liu[2], and Viktor Klep[2]

[1]Department of Chemistry and Biomolecular Science, Clarkson University, Potsdam, NY 13699
[2]School of Materials Science and Engineering, Clemson University, Clemson, SC 29630

The present communication focuses on a novel method for the fabrication of smart coatings based on colloidal systems that combine nanoparticles and spherical responsive polymer brushes. The motivation for this work relies on the traditional and successful applications of composite coatings produced from a polymer matrix and inorganic particles. New properties of such hybrid systems can be approached by the precise design of the dimensions, structure and functionality of the building blocks.

Introduction

As nanoscopic science, technology, and engineering have rapidly emerged in recent years, nanoparticles (colloidal particles in the nanometer size range) have attracted a great deal of attention due to their intriguing electronic, optical, magnetic, transport, mechanical and/or catalytic properties.[1] The unique properties of the particles are generally associated with their nano/quantum-scale dimensions and extremely high surface/volume ratio. Recently, the synthesis and characterization techniques of the nanoparticles have rapidly developed, indicating the importance of organizing those particles into various assemblies to obtain novel properties, which may arise from their 2D and 3D arrangements.[2] Thus, there is a strong need to build an appropriate level of interactions between the neighboring nanoparticles as well as between the nanoparticles and their host environment. To this end, significant advances have been made in the surface modification of the particles by macromolecules. The obtained polymer layers that attach to the particles govern he chemical properties of such particles and their contact interaction with the environment and neighbors, whereas both the particle core and the surrounding organic layer govern their physical properties.[3]

The objective of the reported research was to develop an approach to the fabrication of stimuli responsive colloidal systems (suspensions of smart nanoparticles) and coatings based on the systems. The method employed surface modification of the particles with polymer shells possessing adaptive and responsive behavior. This approach created new types of (smart) nanoparticles capable of reversing the contact interactions between them as well as between the particulates and surrounding media. In this manner, novel responsive/adaptive colloidal systems possessing quasi-stable organization, which can be tuned by an external signal, emerged. The systems are capable of switching interfacial interactions upon an external signal due to the reversible change of the surface chemical composition and the reversible patterning of the surface of the smart nanoparticles. Such nanoparticles are capable of self-assembly into hierarchical structures through noncovalent forces (mesoscale self-assembly [4]) in suspensions and water-oil interfaces. The developed approaches may be translated directly to various colloidal systems containing nanoparticles as an active part for the fabrication of smart coatings.

The major building block for the responsive systems is a mixed polymer brush demonstrating responsive behavior. The term "mixed polymer brush" denotes a thin polymer film constituted from at least two different kinds of immiscible polymer chains randomly grafted (tethered via strong bonds) to the solid substrate.[5] Recent investigations of mixed polymer brushes have demonstrated the possibilities of switching interfacial energy at solid-liquid interface in a broad range (for example, between hydrophilic and ultrahydrophobic behavior).[6] Various outside signals, such as solvents of different solubility parameters, temperature, pH, ionic strength, and ultraviolet light can be used to provide or regulate the switching behavior. To avoid unfavorable interactions, the polymers undergo phase segregation. The chemical grafting to the substrate suppresses a macroscopic segregation. The polymers segregate microscopically into domains that scale with the end-to-end distance of the polymer coils.[7]

The phase segregation mechanism is very sensitive to the environment where a subtle interplay between segregated phases can be shifted by an external signal.[8] We have used this mechanism in stimuli-responsive colloidal systems where the mixed polymer brush is grafted onto the surface of nanoparticles (Figure 1). Switching between different segregated states in the mixed brush allowed us to regulate aggregation of the nanoparticles in a controlled

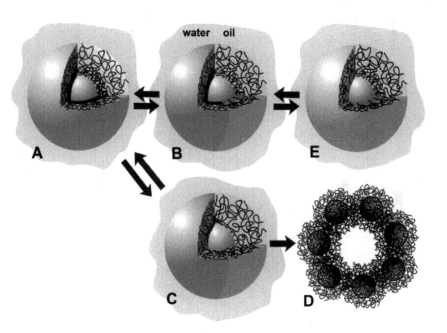

Figure 1. Schematic of the smart particles (mixed spherical brush grafted onto nanoparticles) that undergoes reversible transitions between core-shell morphologies (A, E) and Janus morphology at "water-oil" interface (B) and in nonselective solvent (C). The unsymmetrical morphology results in formation of supermicelles (D). Two unlike polymers are designated as red and blue chains. (See page 2 of color insert.)

environment [9, 10] as well as to regulate the structure of "water-oil" emulsions stabilized or destabilized by the smart particles upon external signals.

Results and Disscussion

We have been developing several approaches to grafting responsive polymer layers onto silica nanoparticles (particle sizes range from 0.02 μm to 0.014 μm). Specifically, the layers can be obtained via consecutive grafting of two polymers and via grafting of block-copolymers.

Synthesis via Consecutive Grafting of Two Polymers

The chemical grafting of polymer brushes can be accomplished by either "grafting to" or "grafting from" methods.[11] According to the "grafting to" technique, end-functionalized polymer molecules react with complementary functional groups located on the surface to form the tethered chains. The "grafting from" technique utilizes the polymerization initiated from the substrate surface by attached (usually by covalent bonds) initiating groups. However, an important question in the fabrication of the grafted layers is how to modify the surface to make it active towards the end-functionalized polymers (for the "grafting to" method) or the initiators (for the "grafting from" technique).

It is necessary to highlight that most of the developed grafting ("to" and "from") methods require attachment of end-functionalized polymers or low molecular weight substances (e.g. initiators) to the substrate for the brush synthesis. There are two common approaches for the attachment of polymerization initiators or end-functionalized polymers in the brush fabrication. The first relies on the reactions between end-functionalized initiator/polymer and native functional groups originally present on the substrate surface.[12] A different approach involves the formation of a monolayer consisting of functional groups active towards terminally functionalized (e.g. epoxide, amine, anhydride or hydroxide) initiator/polymer.[13] Usually the coupling methods are relatively complex and specific for certain substrate/ (macro)molecule combinations. An alternative method for the attachment involves a primary polymer (mono)layer with activity towards both surface and functionalized (macro)molecule.[14] The polymer is used for the initial surface modification as well as for generation of the highly reactive primary layer.

When deposited on a substrate, the primary layer first reacts with the surface through formation of covalent bonds (Figure 2). The reactive units located in the "loops" and "tails" sections of the attached macromolecules are not connected to the surface. These free groups offer a synthetic potential for further chemical modification reactions and serve as reactive sites for the subsequent attachment of the initiators. If the polymer used for building the primary layer contains functional groups highly active in various chemical reactions, the primary layer approach becomes virtually universal towards both particle surface and end-functionalized species being used for the brush formation. In our previous research it was demonstrated that employing poly(glycidyl methacrylate) (PGMA) as a primary anchoring layer, polymer brushes can be successfully synthesized by "grafting from" and "grafting to" techniques on various substrates.[14, 15] We chose the polymer with epoxy functionality since the epoxy groups are quite universal. They can react with different functional groups (carboxy, hydroxy, amino, and anhydride) that are often present or can be created on the surface of various materials. A PGMA layer was successfully deposited on polymeric (PET, polyethylene, silicon resin, nylon) and inorganic (silica, glass, titanium, alumina, gold, silver) surfaces. The epoxy groups of the polymer chemically anchor PGMA to a substrate surface as shown for the silica surface in Figure 2. The glycidyl methacrylate units located in the "loops" and "tails" sections of the attached macromolecules are not connected to the surface. These free groups served as reactive sites for the subsequent attachment of polymer with functional groups or initiators of polymerization, which exhibit an affinity for the epoxy modified surface. Hydrophobic and hydrophilic homopolymers, and statistical and block copolymers were firmly attached to the surface through PGMA.

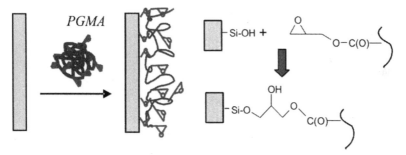

Figure 2. Reactive polymer attached to substrate. (See page 2 of color insert.)

When the mixed, responsive polymer brushes are considered, a variety of parameters influence the mixed brush structure and morphology, such as solubility of the grafted polymers in a solvent, substrate preference of the polymers, grafting density, molecular weight, thickness, and composition. Namely, those parameters dictate the overall switching/responsive performance of the brush. Thus, the availability of the reasonably universal surface modification methods that offer the opportunity to alter the parameters of the brush in a wide range is important. Activation of the surface with a reactive primary layer, in fact, has offered the opportunity to carry out the grafting employing the different grafting approaches. However, the same initial surface modification step is used. We demonstrated that the macromolecular anchoring layer approach allowed fabrication of the mixed polymer brushes by the "grafting to" and "grafting from" methods and their combination. Figure 3 displays the morphology of one of the synthesized mixed brushes. The brushes demonstrated pronounced tendencies to lateral and layered phase segregation.

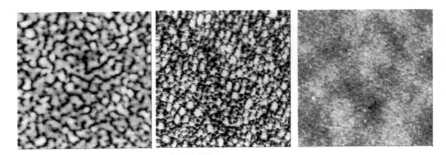

Figure 3. SPM topography images of mixed PS/polyacrylic acid brushes treated with different solvent. From left to right: benzene- THF- ethanol. The brush fabricated by the grafting to/grafting from combination. Image: 1x1 μm

In order to conduct detailed studies of the deposition of the anchoring layer on silica nanoparticles (diameter 150 nm), we synthesized a special version of an epoxy-containing polymer PGMA. Specifically, the polymer was labeled with the fluorescent dye, Rhodamine B. The labeling allows the level of deposition of the polymer on the nanoparticle surface to be controlled. Indeed, the PGMA-modified nanoparticles demonstrated a strong fluorescence signal, confirming the surface modification. The labeling of the particles is also important for studying particle aggregation and migration from one solvent to another. In addition, labeling provides an easy means of studying the adsorption of particles on different surfaces. The grafting performance of the labeled epoxy-containing polymer was comparable to that of the unlabeled one. A poly(2-vinylpyridine) (P2VP)/polyethylene glycol (PEG) mixed brush was successfully grafted to the nanoparticle activated using the fluorescent-labeled PGMA.

Synthesis via Grafting of Block-Copolymers

This method uses functional groups of a block-copolymer to graft the copolymer onto the surface of silica particles (Figure 4). Because of the statistical character of the grafting in this case, the grafting point of the block-copolymer on the particle surface may be located in different distances from the grafting point between two blocks. This results in the formation of a polydispersed mixed brush. The first step was the functionalization of the surface of silica particles with 11-bromoundodeciltrimethoxisilane (BTMS). The second step was the grafting of the block-copolymer: poly(styrene-b-4-vinyl pyridine) P(S-b-4VP) with the molecular masses of the blocks: PS $Mn=20,000$ g/mol, P4VP $Mn=19,000$ g/mol, $PI=1.09$. The grafting was performed via quaternization reaction between the bromoalkyl functionalities on the particle surface and P4VP. The grafted amount of the block-copolymer on the silica particles of 190 nm in diameter was 3.8 mg/m2 (measured using elemental analysis).

Fabrication and Study of Smart Coatings

The particles with the mixed brush shell (hybrid particles) were used to fabricate smart responsive coatings. Silica wafers were modified with BTMS by the same method as the silica particles, and then the particles were grafted onto the silica wafers by quaternization reaction (Figure 5).

The deposition of nanoparticles on silica wafers was performed using the vertical deposition approach,[16] where solvent evaporation provides the transport of the particles to the meniscus and the formation of ordered arrays. We prepared three percent suspension of the synthesized particles in dichloromethane. Then silica wafer was immersed in the solution and the solvent was evaporated under low pressure.

The data of contact angle measurements demonstrated well pronounced switching from hydrophobic to hydrophilic wetting behavior upon exposure to different solvents. For example, the treatment of the sample with toluene resulted in a 125° water contact angle, while rinsing with acidic water of pH 2 switched the sample to a hydrophilic state (water contact angle 44°).

Smart Coatings with Complex Texture. Ultrahydrophobic Coatings

We used the responsive behavior of the hybrid particles to regulate the sizes of aggregates formed by the particles. We found that the aggregation of the particles can be tuned via pH change or by adding organic solvents to aqueous dispersions of the particles.

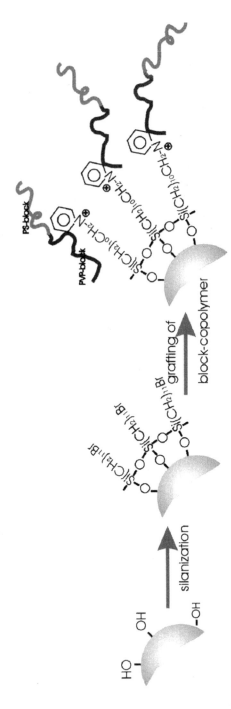

Figure 4. Schematic representation of the grafting route to fabricate mixed brushes from block-copolymer on the silica surface: step 1-chemosorption of BTMS; step 2-grafting of the diblock copolymer via quaternization of the poly(4-vinylpyridine) block. (See page 3 of color insert.)

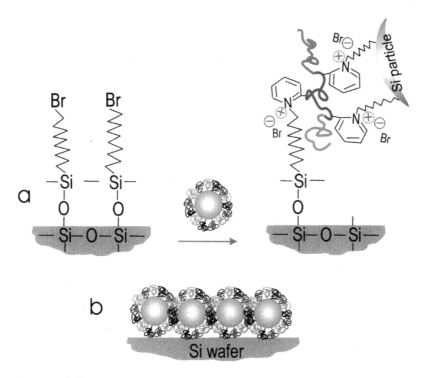

Figure 5. Schematic representation of grafting silica nanoparticles with block copolymer brushes to the Si wafer surface (a); a layer of the nanoparticles on Si wafer (b)

The particles were dispersed in toluene and an acidic aqueous environment. In both solvents, dynamic light scattering and AFM investigations concluded that they formed stable suspensions. The stability of the particles' suspensions in different media suggests that the particles change and adjust their interfacial energy in the suspensions. The particles' shell provides versatile surface transitions between different core-shell morphologies in selective solvents as depicted in Figure 1. In water at pH 2, protonated P4VP chains are exposed to the particles' exterior while PS is segregated to the inner part of the brush (Figure 1A) and vice versa in toluene (Figure 1E). An increase in the pH of the particles' water suspension increases the effect of hydrophobic interactions due to the deprotonation of P4VP, resulting in aggregation of the particles. The aggregation is a reversible process and the aggregation-redispersion can be reversed by alternating pH values between the regimes of stability and aggregation. If water is gently added to the dispersion of the particles in toluene, the particles populate the toluene water interface. The addition of toluene to an aqueous dispersion of the particles leads to their spontaneous adsorption at the water-toluene interface, and leads as well to switching the morphology of the particles' shell (Figure1B). Thus, on the particle site in contact with the toluene phase, PS blocks are expanded in the toluene phase, whereas P4VP are collapsed, and the reverse is true for the particle site next to the water phase.

Using the particles' aqueous dispersions and their water-toluene emulsions, we prepared ultrahydrophobic coatings on the originally-hydrophilic surface of a polyamide reference textile sample. The coatings were deposited on the textile by casting from the particles' suspensions of different pH values. In the aqueous environment, the P4VP shell of the hybrid particles is exposed to the solution and the shell is responsible for a rapid adsorption of the particles on the fabric surface. Wettability of the textiles was investigated by contact angle measurements. After the deposition of the particles, the textile surface remained hydrophilic because the particles' surface was enriched by the hydrophilic polymer blocks. Heating the surface above the glass transition temperature of PS (Tg=100°C) switched the textile surface into the ultrahydrophobic state. As a more hydrophobic polymer, PS is preferentially exposed to the exterior of the particle shell upon heating in air. In this way, the textile surface was switched into the ultrahydrophobic state. The coating can be easily switched back to a hydrophilic state upon exposure to acidic (pH3) water, and hydrophobized again by heating in air or by exposure to toluene. As can be seen from Table 1, the coatings from 15 nm hybrid particles approach the superhydrophobic behavior in the pH range from 2 to 5. The coatings prepared from the particles' water dispersions of pH>5.5 did not produce ultarhudrophobic surfaces.

The same procedure was used to test the coatings prepared from the dispersions of the hybrid particles of 200nm in diameter with added toluene as an alternative way to regulate the size of aggregates. We selected a case in which we observed very few single particles in the emulsions, while a major fraction of particles was located at the water-toluene interface. We observed very pronounced differences in the wetting behavior of annealed coatings obtained from the emulsions of pH values from 2.0 to 10.2 (Figure 6).

Only the coating obtained from the emulsion of pH 3 showed ultrahydrophobic behavior (Figure 6b). This result demonstrated that the ultrahydrophobic properties could be approached if an appropriate surface texture is formed on the sample. Tuning the size of aggregates by the change of pH range of aggregates at pH 3, and the combination of the deposited aggregates with the textured structure of the fabric, allows for the fabrication of appropriate surface morphology for the ultrahydrophobic coatings.

Thus, the responsive nanoparticles demonstrated the dual effect on the coating formation: 1) tuneable sizes of the deposited aggregates, and 2) switching from hydrophilic surface in water to the hydrophobic surface in air upon heating.

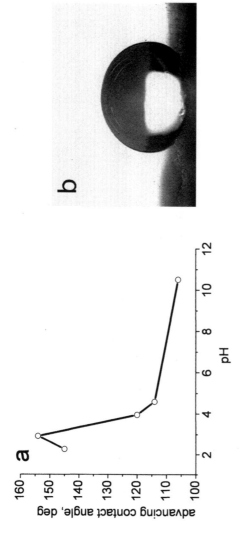

Figure 6. Advancing water contact angle on the textile coated by hybrid particles vs. pH of the water-toluene dispersions (a), and a photograph of the water droplet on the textile coated with the responsive particles from their water-toluene emulsion of pH 3 after annealing at 100 °C (b).

Table 1. Water contact angles on the textile samples coated
with aggregates of the hybrid particles (15 nm in diameter).

pH	2	2.9	4.0	4.9	5.5	6.5
Contact angle, deg	152.9	153.7	152.3	152.6	H*	H*
Roll off angle, deg	<10	<10	<10	<10		

H*- hydrophilic surface, water droplets soaked into the textile.

Conclusions

The general purpose of this work was to create responsive shells on the surface of nanoparticles and use the responsive hybrid particles as building blocks for the fabrication of smart coatings on various solid substrates. The hybrid particles exhibit multiple functions: 1) the responsive shell stabilizes the particle dispersion in aqueous environment; 2) the shell is used to tune and regulate the size of the particle aggregates to approach a specific texture of the deposited coatings, which will result in the ultrahydrophobic effect, 3) the responsive shell is capable of building a strong interaction with the substrates; and 4) the shell switches from the hydrophilic surface of the particles to a hydrophobic surface upon the drying of the coatings on the substrate, and vice versa upon treatment with appropriate solvents.

Acknowledgment

We acknowledge financial support from the NSF awards CTS 0456548 and CTS 0456550, and NTC award CO4-CL06.

References

1. Ohno, K.; Koh, K.; Tsujii, Y.; Fukuda, T. *Macromolecules* **2002**, *35*, 8989.
2. Ohno, K.; Koh, K.; Tsujii, Y.; Fukuda, T. *Angew. Chem., Int. Ed.* **2003**, *42*, 2751.
3. Mori, H.; Seng, D. C.; Zhang, M. F.; Muller, A. H. E. *Langmuir* **2002**, *18*, 3682.
4. Bowden, N. B.; Weck, M.; Choi, I. S.; Whitesides, G. M. *Acc. Chem. Res.* **2001**, *34*, 231.

5. Minko, S.; Müller, M.; Luchnikov, V.; Motornov, M.; Usov, D.; Ionov, L.; Stamm, M. In *Polymer Brushes*. Advincula, R. C.; Brittain, W. J.; Caster, K. C.; Ruehe, J., Eds. Wiley-VCH: Weinheim, 2004; p 403.
6. Luzinov, I.; Minko, S.; Tsukruk, V. V. *Prog. Polym. Sci.* **2004**, *29*, 635.
7. Müller, M. *Phys Rev E* **2002**, *65*, 30802.
8. Minko, S.; Muller, M.; Usov, D.; Scholl, A.; Froeck, C.; Stamm, M. *Phys. Rev. Lett.* **2002**, *88*, 035502.
9. Singh, C.; Pickett, G. T.; Zhulina, E.; Balazs, A. C. *J. Phys. Chem. B* **1997**, *101*, 10614.
10. Chem, S. S.; Zhulina, E. B.; Pickett, G. T.; Balazs, A. C. *J. Chem. Phys.* **1998**, *108*, 5981.
11. Zhao, B.; Brittain, W. J. *Prog. Polym. Sci.* **2000**, *25*, 677.
12. Jones, D. M.; Brown, A. A.; Huck, W. T. S. *Langmuir* **2002**, *18*, 1265.
13. Luzinov, I.; Julthongpiput, D.; Malz, H.; Pionteck, J.; Tsukruk, V. *Macromolecules* **2000**, *33*, 1043.
14. Iyer, K.; Zdyrko, B.; Malz, H.; Pionteck, J.; Luzinov, I. *Macromolecules* **2003**, *36*, 6519.
15. Liu, Y.; Klep, V.; Zdyrko, B.; Luzinov, I. *Langmuir* **2004**, *20*, 6710.
16. Jiang, P.; Bertone, J. F.; Hwang, K. S.; Colvin, V. L. *Chem. Mater.* **1999**, *11*, 2132.

Chapter 8

Nanostructured Electrooptically Active Smart Coatings Based on Conjugated Polymer Networks: Precursor Polymer Approach, Devices, and Nanopatterning

Rigoberto C. Advincula

Departments of Chemistry and Chemical Engineering, University of Houston, Houston, TX 77204

In this paper, we report several strategies for preparing smart coating based on the synthesis, electrodeposition, and patterning of ultrathin films of conjugated polymer network (CPN) films on flat electrode surfaces using the "precursor polymer" approach. This involves a rational synthesis and design of a precursor polymers followed by careful electrodeposition and characterization of ultrathin films and coatings on conducting substrates. A precursor design involves the tethering of electropolymerizable monomers to polymer backbones that can have "active" or "passive" electro-optical properties. This has resulted in the preparation of smooth, high optical quality films, which should be important for practical electro-optical coatings applications. Characterization of these films has been made using surface sensitive spectroscopic and microscopic analytical techniques. Copoly-merization with monomers, polymer backbone design, and grafting on modified surfaces are key points in a coatings strategy. Novel methods of in-situ characterization techniques have also been developed combing electrochemistry and surface plasmon resonance techniques. These coatings have found applications for devices as well as sensors.

Introduction

Research on conducting and π-conjugated polymers utilized for organic semiconductor and display materials is an interdisciplinary approach.*(1)* Chemists, physicists, and engineers have investigated their synthesis, properties, and processing over the past decades, both for fundamental reasons and for their potential applications. These polymers are often intractable and hard to process but they have been made soluble by choosing proper substituents (usually alkyl groups) that are incorporated in their design and synthesis. The most well-studied polymers include: polyanilines, polypyrroles, polythiophenes, polyfluorenes, etc.[1] These polymers can undergo striking physical or chemical changes when exposed to heat, light, electric fields and various chemical moieties (dopants) giving rise to conductivity, fluorescence, reversible thermochromism, photochromism, electrochromism, and even ionochromism. Recently, the synthesis of conjugated dendrimers has been reported by several groups including our group.*(2)* These conjugated polymer dendrimers are often more soluble than their linear derivatives. They can be prepared through a number of metal-mediated cross-coupling reactions based on conjugated dendron structures that can be linked to a core group. In addition, conjugated polymers can be blended with nanoparticle materials to show improved photophysical and photochemical properties related to energy transfer and electron transport properties.*(3)* We have recently investigated the preparation of hybrid thiophene dendron-nanoparticle materials.*(4)* Thus they have found applications as conducting polymers *(1)*, polymer light emitting diode (PLED),*(5)* field-effect transistor (FET) devices,*(6)* and even lasers.

We have reported in a number of papers a novel method for depositing high optical quality ultrathin coatings of conjugated polymers on conducting substrates.*(7)* Using a "precursor polymer" route we have employed electro-deposition as a method to prepare ultrathin films on conducting substrates. We have demonstrated the viability of the "precursor polymer" approach based on single pendant electroactive monomer group using various polymer backbones (Figure 1). This involved the tethering of electroactive monomers to a polymer backbone that can be labeled as "passive" or "active" in terms of its contribution to the electro-optical properties of the film itself. Thus by design the pendant electroactive monomer groups undergo electropolymerization or chemical oxidation resulting in a conjugated polymer network (CPN) film or coating. The network formation is based on both *inter-* and *intra-*molecular cross-linkages occurring between the pendant monomer units, i.e. of the same chain or a different chain. The films formed are characterized by high optical quality (transparency), uniform coverage, good adhesion, smoothness in morphology, and controlled ion permeability. Moreover, by controlling the amount of conjugated species and doping, it should be possible to control electrical conductivity. Thus, the process is interesting for depositing *insoluble* cross-

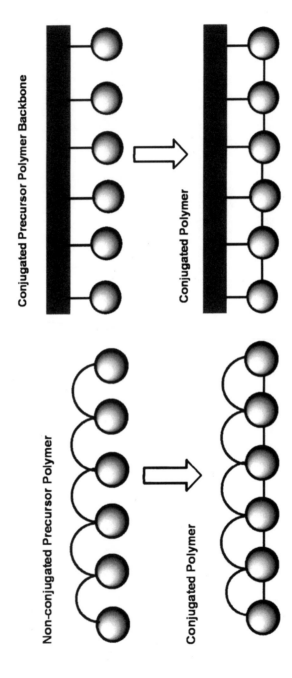

Figure 1. The Precursor Polymer approach to conjugated polymers with different polymeric backbones: passive and active.

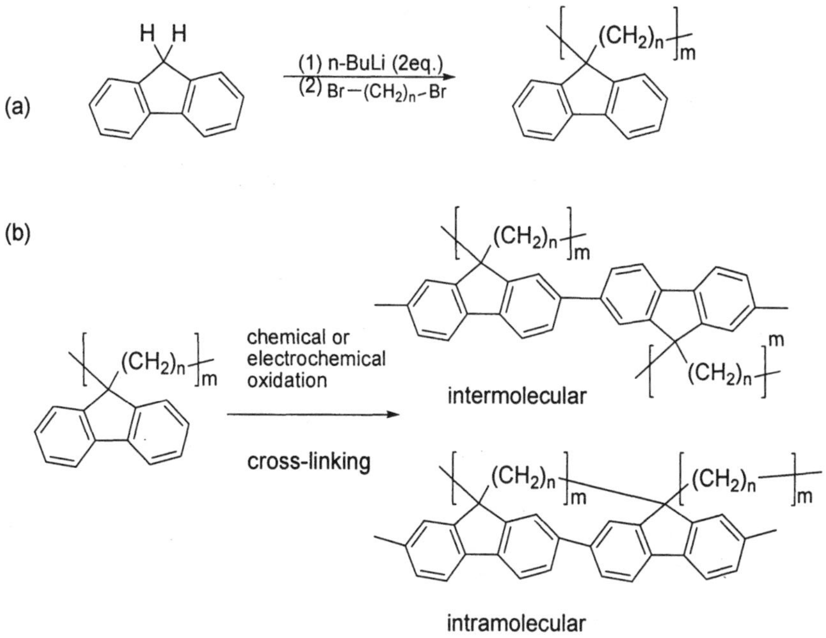

Figure 2. Crosslinking properties of precursor polyfluorenes: a) Synthesis of fluorene-containing polymers with n = 4,6,8,10,12. (b) Chemical or electrochemical oxidation leading to the formation of cross-linked or a conjugated polymer network. (Reproduced from reference 9)

linked ultrathin coatings of conjugated polymers for practical electro-optical applications. A variety of combinations should be possible for the design of a precursor polymer backbone and the "electroactive monomer" side group to form the CPN film. It is also possible to *copolymerize* with small molecule electroactive monomers with varying compositions in order to control the degree of cross-linking and linear polymer formation. This paper reports these various designs, combinations, and methods of coating preparation and patterning protocols.

Results and Discussion

Polyfluorene Precursor Polymers

The precursor polymer approach was first demonstrated by our group on a series of poly(fluoren-9,9-diyl-*alt*-alkan-diyl) polymers, which was converted to conjugated poly(2,7-fluorene)s both by chemical and electrochemical oxidation into conjugated poly(2,7-fluorene) derivatives (Figure 2).*(8)* These polyfluorene precursor polymers have interesting design properties in which the distance between adjacent fluorene groups can be varied by an alkylene spacer unit. This affected the thermal properties of the polymers where the Tg shifted to lower values with increasing spacer length. In collaboration with Agilent Technologies, PLED devices were prepared by depositing electrochemically on selected areas of a substrate.*(9)* Efficient devices were prepared based on different hole-transport layers. The effectiveness of this experiment demonstrated the applicability of this technique to "site directing" and patterning. We also reported a unique polyfluorene precursor structure in which the "pre-formed" conjugated polyfluorene has pendant carbazole units which were found to be capable of forming robust film structures by "electrografting" on ITO surfaces.*(10)* The ITO surface was first functionalized with a self-assembled monolayer (SAM) of silane-carbazole derivatives. The CPN films were then prepare either from solution or by first spin-coating on the ITO substrates. These films showed improved fluorescence properties based on a decrease of exciplex formation due to chain-to-chain aggregation in polyfluorenes. Since then, we have extended this concept to new methods of synthesis, electrodeposition, characterization, patterning, and grafting of conjugated polymers on planar electrode surfaces as we describe in this paper. While a number of homopolymers and copolymers have been reported which contain one type of tethered monomer, we have also recently reported the preparation of binary tethered monomer compositions pendant on a polymer backbone.*(11)*

Polymethylsiloxane Precursor Polythiophenes and Polypyrroles

We have described the synthesis, electropolymerization, and ultrathin film formation of polymethylsiloxane modified polythiophene and polypyrrole precursor polymers.*(12)* We have used the unique properties of the silicone polymers towards depositing ultrathin films of conjugated polymers with *high optical quality* on surfaces. Like the previous work on polyfluorenes, the mechanism of electropolymerization has differences in parameters with solution polymerization involving small monomeric units. In terms of film formation, our emphasis was on determining the unique parameters affecting film formation, optical quality (refractive index and thickness, dielectric constants), morphology, etc. in order to optimize for specific electrical or luminescence properties. The polydimethylsiloxane backbone is a good polymer backbone because of its low Tg, inertness, low reactivity, and property contrast with other polyolefinic and methacrylate polymers. In essence, the high optical clarity of silicone polymers was an advantage for coatings formation. The synthesis of the polymers was done by a hydrosilylation route with polyhydroxymethylsiloxane (Figure 3).

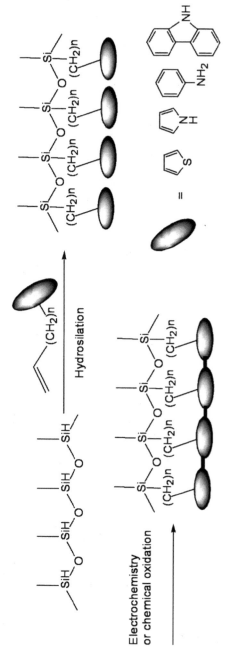

Figure 3. A general illustration of the polymethylsiloxane modified precursor polymer route with other electroactive monomers. (Reproduced from reference 12)

These reactions were followed by ^1H NMR and ^{13}C NMR and elemental analysis. The precursor siloxane polymer was cross-linked by electropolymerization and monitored by cyclic voltammetry at various solvents, substrates, solution concentrations, and counter electrolytes. The electropolymerization was performed by sweeping the potential typically from $-$ 400mV to 1500mV vs Ag/Ag$^+$ reference electrode (Figure 4). After several cycles, the substrate was taken out and rinsed with solvent, dried under dry nitrogen flow and analyzed.

The precursor polymer can be crosslinked through the 2,5 linkages of thiophene by electrochemistry or oxidative chemical methods. The films are very uniform and are optically clear. This uniform crosslinked polymer film is essentially *insoluble* to solvents. The films also showed reversible color changes (electrochromism) during the redox process. For example, the precursor polythiophene turns to deep purple upon oxidation, and then changes back to orange during the corresponding reduction process. Different morphologies have been observed by AFM imaging on films at different cycling stages. Films were also prepared by copolymerizing with different compositions of the thiophene monomer.

For the pyrrole grafted to the polymethylhydrosiloxane to form poly(methyl-(10-pyrrol-3-yl-decyl)siloxane,*(13)* interesting morphologies and optical behavior were also observed (Figure 5). This consisted of circular "nano-domains", 200-300 nm, which we believe to be phase separated polysiloxane rich regions on the surface. By copolymerizing with pyrrole monomers at different composition ratios, these domains vary in size and distribution as a function of composition. The domain size decreased while roughness was found to increase with increasing amount of pyrrole comonomer content. It will be interesting to compare the electrical conductivity of the resulting cross-linked polymers as a function of composition and correlation with morphology.

PMMA, PVK, and a Substituted Polyacetylene Polymer Backbone

We have also prepared pyrrole grafted on a polymethylmethacrylate (PMMA) polymer backbone by first synthesizing the monomer (N-alkyl pyrrole methacrylate ester) followed by polymerization with AIBN to form poly 6-(N-pyrrolyl)hexyl methacrylate **(PHMA)1** (Figure 6) *(14)* Likewise, by cyclic voltammetry (100 mV/s), electropolymerization and cross-linking was done at different solution conditions on a flat ITO and Au electrodes for these polymers. Copolymerization with other pyrrole monomers at different composition ratios was also done. Studies are currently being made to understand the composition dependence of the different morphologies observed and the electrical and optical properties of these films.

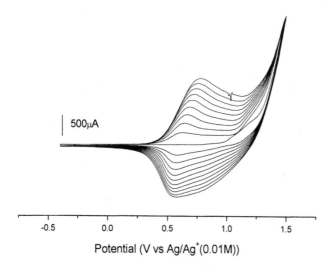

Figure 4. A typical CV for the poly(methyl-(10-thiophen-3-yl-decyl)siloxane precursor polymer. Conditions: Three-electrode cell, TBAH as electrolyte, and scan rate of 100mV/s. (Reproduced from reference 12. Copyright 2001 American Chemical Society.)

Substituted Polyacetylenes with cross-linked carbazole units

Poly(N-vinylcarbazole) (PVK), which exhibits interesting electrical properties, has been studied extensively because of its possible applications in light emitting diodes, polymer batteries, and for applications in various electrochromic devices.[15],[16] Polyacetylene is the first reported conductive polymer with both high conductivity and interesting electro-optical properties.[17] In this work, we described the synthesis and electropolymerization of conjugated substituted polyacetylenes, poly(N-alkoxy-(p-ethynylphenyl)carbazole) or **PPA-Cz-Cn** with electropolymerizable carbazole side groups to form conjugated polymer network (CPN) films (Figure 7a). The phenylacetylene monomer was functionalized with a carbazole group separated by an alkylene spacer. Polymerization of the monomer in solution is accomplished using a Rh catalyst to form a "precursor polymer". The electrochemical behavior and cross-linking of the carbazole side group was then investigated by cyclic voltammetry (CV) and spectroelectrochemistry (Figure 7b). A trend in the redox electrochemical behavior was observed with varying alkyl spacer length between the polyphenylacetylene backbone and carbazole side-group. The resulting film combines the electro-optical properties of the conjugated polyphenylacetylene polymer with polycarbazole units in a cross-linked electropolymerized film as evidenced by the CV and spectroelectro-

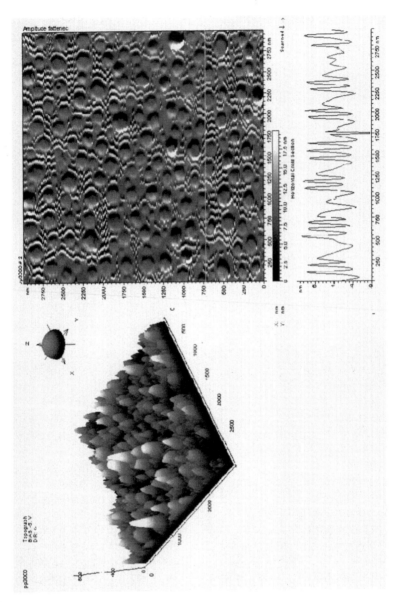

Figure 5. Morphology of the poly(methyl-(10-pyrrol-3-yl-decyl)siloxane film by AFM imaging using a magnetic-AC (MAC) mode. (Reproduced from reference 13.)

167

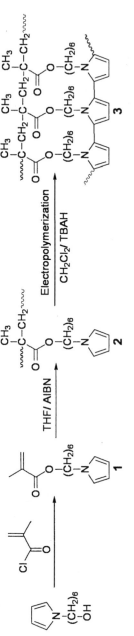

Figure 6. Selective electropolymerization of the functional groups in poly 6-(N-pyrrolyl)hexyl methacrylate (PHMA) 1 to form a conjugated polymer. (Reproduced from reference 14)

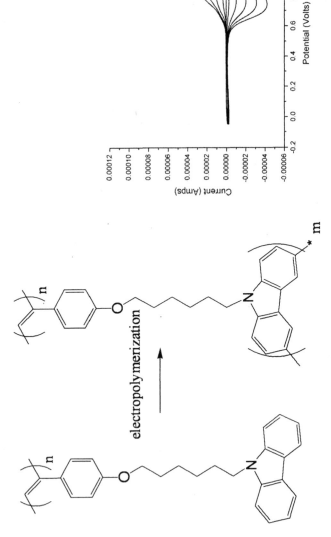

Figure 7. Electropolymerization and crosslinking scheme for carbazole substituted polyacetylene:
(a) structure of the polyacetylene-carbazole precursor and the reactivity of the carbazole units.
(b) CV deposition cycle of the precursor polymer showing the formation of lower oxidation potentials
with subsequent cycles. (Reproduced from reference 11. Copyright 2005 American Chemical Society.)

chemical behavior. Thus, this study emphasized the preparation of polymer materials with mixed π-conjugated species arising from the electrochemical cross-linking of a designed precursor polymer.

Electrodeposition of Polycarbazole Precursors on Electrode Surfaces and Devices

The electrochemical doping properties and morphology changes of electropolymerized poly(N-vinylcarbazole) (PVK) thin films were investigated towards improved hole-injection/transport properties in polymer light emitting diode (PLED) devices (Figure 8).*(18)* The conjugated network polycarbazole thin films (resulting from a poly(N-carbazole) (PCz) network) were prepared by electrodeposition of PVK and/or N-vinylcarbazole co-monomer via precursor route and were investigated *in situ* by electrochemical-surface plasmon resonance spectroscopy (EC-SPS). Distinct doping-dedoping properties and morphology transitions were observed with different compositions of PVK and Cz. By electrochemical doping of the cross-linked conjugated polycarbazole units, the electrochemical equilibrium potential (E_{eq}), which correlates to Fermi level (E_f) or the work function (ϕ_w) of the film, was adjusted in the vicinity of the anode electrode (Figure 9). The conjugated network PCz films were then used as a hole-injection/transport layer in a two layer device. Remarkable enhancement of PLED properties was observed when optimal electrochemical doping was done with the films. Important insight was gained on charge transport phenomena between polymer materials and conducting oxide substrates.

Recently, we have also reported the synthesis and electropolymerization of a precursor polymer with a binary molecular composition of thiophene and carbazole electroactive groups to form ultrathin films of conjugated polymer networks (CPN) on flat indium tin oxide (ITO) substrates.*(11)* While in the past we have demonstrated the precursor polymer approach based on a single pendant electroactive group, in this work, we describe the interesting electro-copolymerization mechanism and properties of precursor polymers prepared with two different types of pendant electroactive groups (statistical copolymer) and compared behavior to their respective homopolymers. The presence of a smaller amount of carbazole induces the electropolymerization of the higher oxidation potential thiophene units via the reaction of a radical cation and a neutral molecule pathway. These electrochemically generated thin films gave unique optical, electrochemical and morphological properties as a function of composition. Thus the mechanism involved a "trigger" effect for the thiophenes based on the initial oxidation of the carbazole units in a radical cation to neutral monomer reaction pathway. Both copolymer and homopolymer electropolymerizations resulted in statistical intra-molecular and inter-molecular reaction between individual polymer chains resulting in templating and cross-

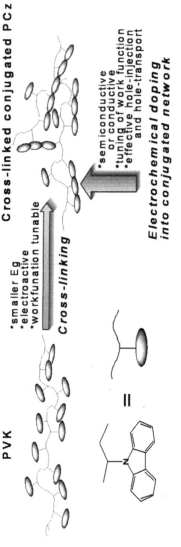

Figure 8. A strategy to use the precursor PVK route as both hole-injection layer and hole-transport layer. PVK is crosslinked to form oligocarbazole units.

171

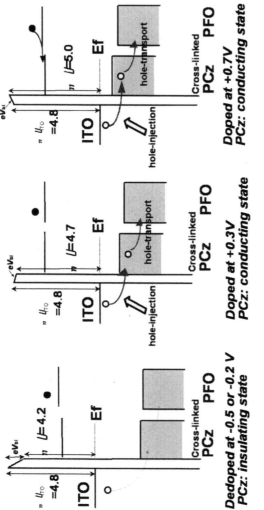

Figure 9. Schematic drawing of the proposed mechanism of the hole-injection/transport and the various relative energy levels. (Reproduced from reference 18)

linking. This was verified by CV, spectroelectrochemistry, and XPS, measurements. The morphology of the films correlated well with the deposition behavior.

Patterning Strategies:

Micro-contact printing is a soft-lithography approach that has been used for creating micron sized patterns on flat surfaces.*(19)* By using an elastomeric stamp prepared from a photolithographically prepared master, various size resolutions and patterns have been prepared using an amphiphilic inks capable of forming self-assembled monolayer (SAM)s on a variety of surfaces. The pattern fidelity relies on differences in surface energies and specific amphiphile to substrate interactions (chemical bond).

In a recent paper, have used selective area electropolymerization to generate patterned conjugated and fluorescent polymer thin films.*(20)* Instead of using electroactive monomers, we have used a precursor polyfluorene polymer, which has fluorescent properties defined by the polymer main chain and electropolymerizability of pendant carbazole units. The polymer preferably deposited on the alkyl SAM defined regions to form sharp patterns. The patterns have been characterized by atomic force microscopy (AFM). The fluorescence micrograph confirmed that the polymer main chain was largely unaffected by the redox electrochemistry and it also showed highly regular patterning characteristics in a large area (Figure 10). This new combination of site selective electropolymerization and the precursor approach may provide a new way to make patterned polymer light-emitting diodes in the future.

The direct nanopatterning under ambient conditions to form conjugated polycarbazole patterns from ultrathin films of a "precursor polymer" and monomer has been recently reported by us.*(21)* In contrast to previous reports on electrochemical dip-pen nanolithography using monomer ink or electrolyte-saturated films in electrostatic nanolithography,*(22)* these features were directly patterned on spin-casted carbazole monomer and polyvinyl carbazole (PVK) polymer films under room temperature and humidity conditions. The nanopatterned electric field using a biased atomic force microscope (AFM) tip induced polymerization and cross-linking between carbazole units in the films (Figure 11). Different parameters including writing speed and bias voltages were optimized to control line width and patterning geometry (Figure 12). The conducting property (I-V curves) of these nanopatterns was also investigated using the conducting- AFM (C-AFM) setup and the thermal stability of the patterns were evaluated by annealing the polymer/monomer film above the glass transition (T_g) temperature of the polymer. To the best of our knowledge, this is the first report in which *thermally stable* conducting nanopatterns were drawn directly on monomer or polymer film substrates using an electrochemical

Figure 10. Fluorescence microscopy of the non-lithographically patterned film made under potentiostatic conditions showing high pattern fidelity. The structure of the polyfluorene copolymer is also shown. (Reproduced from reference 20. Copyright 2004.)

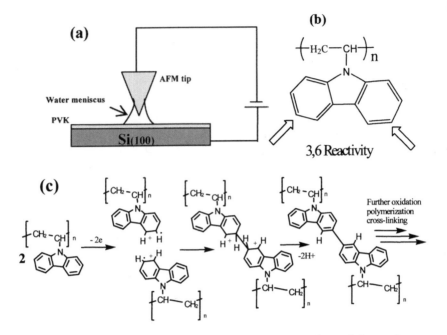

Figure 11. (a) Schematic diagram of electrochemical nanolithography
(b) chemical structure and possible polymerization sites of PVK
(c) mechanism for electropolymerization (cationic) and cross-linking of PVK.
(Reproduced from reference 21. Copyright 2006 American Chemical Society.)

nanolithography technique under ambient conditions. Thus, this study will open up the investigation of other precursor polymer materials, electropolymerizable monomers, and ultrathin film assemblies with parameters of patterning relating, to Tg of the polymer, electropolymerizatibility, writing speed and applied potential. A main advantage of this method is the use of ambient temperature and humidity conditions for these experiments.

Conclusion

We have developed strategies for electrodepositing and patterning ultra-thin films of conjugated polymers on flat electrode surfaces using the precursor polymer approach. This has resulted in the preparation of smooth, high optical quality films, which should be important for applications involving flat electrode surfaces in devices. Copolymerization with monomers, polymer backbone design, and grafting on modified surfaces are key points in this

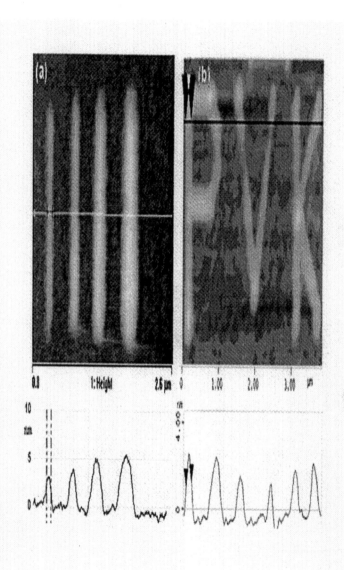

Figure 12. (a) Nanolines written on PVK film at constant tip speed of 1 m/s with the different bias of -5V, -7V, -9V and corresponds to the line width of 83nm, 128nm, 162nm, and 231 nm, respectively, (b) AFM image with height profile of the pattern "PVK" drawn at -7V at a tip speed 1 m/s. (Reproduced from reference 21. Copyright 2006 American Chemical Society.)

strategy. Novel methods of in-situ characterization techniques have also been used.

Acknowledgement

We gratefully acknowledge partial support from NSF-DMR (05-04435), NSF-DMR-(06-02896), and NSF-CTS (0330127), ACS-PRF 45853-AC7, the Robert A. Welch Foundation, the Alexander von Humboldt Foundation and collaborations with Daniel Roitman and Seiji Inaoka of Agilent Technologies, S.A. Karim and Toshio Masuda of Kyoto University, Suresh Valiyaveettil and S. Jegadessan of National University of Singapore, Sukon Phanichphant of Chiang Mai University, and Wolfgang Knoll of the Max Planck Institute for Polymer Research. The Post-doc and students who have been involved with this research include: Akira Baba, Ken Onishi, Sieji Inaoka, and students; Chuangjun Xia, Suxiang Deng, Prasad Taranekar, Mi-Kyoung Park, Derek Patton, Chengyu Huang, Guoqian Jiang, Ramakrishna Ponnapati, Yushin Park, JinYoung Park, and Paralee Waenkaew.

References

1. Salaneck, W.; Lundstrom, I.; Ranby, B. *Conjugated Polymers and Related Materials*; Eds.; Oxford University Press: Oxford, 1993.
2. (a) Xia, C.; Fan, X.; Locklin, J.; Advincula, R. C.; Gies, A.; Nonidez, W. J. *Am. Chem. Soc.* **2004**, *126*, 8735-8743. (b) Xia, C.; Fan, X.; Locklin, J.; Advincula, R. C. *Org. Lett.* **2002**, *4*, 2067-2070. (c) Moore, J. S. *Acc. Chem. Res.* **1997**, *30*, 402-413. (d) Gong, L.; Hu, Q.; Pu, L. *J. Org. Chem.* **2001**, *66*, 2358-2367. (e) Deb, S. K.; Maddux, T. M.; Yu, L. *J. Am. Chem. Soc.* **1997**, *119*, 9079-9080. (f) Meier, H.; Lehmann, M. *Angew. Chem., Int. Ed.* **1998**, *37*, 643-645. (g) Lupton, J. M.; Samuel, I. D. W.; Beavington, R.; Burn, P. L.; Bassler, H. *Adv. Mater.* **2001**, *13*, 258-261. (h) Berresheim, A. J.; Muller, M.; Mullen, K. *Chem. Rev.* **1999**, *99*, 1747-1786.
3. Mattoussi, H.; Radzilowski, L. H.; Dabbousi, B. O.; Thomas, E. L.; Bawendi, M. G.; Rubner, M. F. *J. Appl. Phys.* **1998**, *83*, 7965.
4. Locklin, J.; Patton, D.; Deng, S.; Baba, A.; Millan, M.; Advincula, R. C. Chem. Mater.; (Article); 2004; 16(24); 5187-5193.
5. Segura, J. *Acta Polym.* **1998**, *49*, 319.
6. Bao, Z.; Feng, Y.; Dodabalapur, A.; Raju, V.R.; Lovinger, A.J. *Chem. Mater.* **1997**, *9*, 1299.
7. Inaoka, S.; Roitman, D.; Advincula, R. in *Forefront of Lithographic*

Materials Research, Ito, H., Khojasteh, M., Li, W., Eds. Kluwer Academic Publishers, New York, NY, 2001, p. 239-245.

8. Inaoka, S.; Advincula, R. *Macromolecules* **2002**, *35*, 2426-2428.
9. Inaoka, S.; Roitman, D. B.; Advincula, R. C. *Chem. Mater* **2005**, *17*, 6781-6789.
10. Xia, C.; Advincula, R. C. *Chem. Mater.* **2001**, *13*, 1682-1691.
11. Taranekar, P.; Baba, A.; Fulghum, T. M.; Advincula, R. *Macromolecules* **2005**, *38*, 3679-3687.
12. Xia, C.; Fan, X.; Park, M.-k.; Advincula, R. C. *Langmuir* **2001**, *17*, 7893-7898.
13. Taranekar, P.; Fan, X.; Advincula, R. *Langmuir* **2002**, *18*, 7943-7952.
14. Deng, S.; Advincula, R. C. *Chem. Mater.* **2002**, *14*, 4073-4080.
15. Burrows, P. E.; Forrest, S. R.; Sibley, S. P.; Thompson, M. E. *Appl. Phys. Lett.* **1996**, *69*, 2959.
16. Tamada, M.; Omichi, H.; Okui, N. *Thin Solid Films* **1995**, *268*, 18.
17. Masuda, T.; Higashimura, T. *Acc. Chem. Res.* **1984**, *17*, 51-56.
18. Baba, A.; Onishi, K.; Knoll, W.; Advincula, R. C. *J. Phys. Chem. B.* **2004**, *108*, 18949-18955.
19. Kumar, A.; Biebuyck, H. A.; Whitesides, G. M. *Langmuir* **1994**, *10*, 1498.
20. Xia, C.; Advincula, R. C.; Baba, A.; Knoll, W. *Chem. Mater.* **2004**, *16*, 2852-2856.
21. (a) Jegadesan, S.; Advincula, R. C. Valiyaveettil, S. *Adv. Mater.* **2005**, *17*, 1282-1285. (b) Jegadesan, S.; Sindhu, S.; Advincula, R. C.; Valiyaveettil, S. *Langmuir* **2006**, *22*, 780-786.
22. (a) Lim, J. H.; Mirkin, C. A. *Adv. Mater.* **2002**, *14*, 1474-1477. (b) Piner, R. D.; Mirkin, C. A. *Langmuir* **1997**, *13*, 6864-6868.

Chapter 9

Synthesis of Acrylic Polymer Nanoparticles: Speculation about Their Properties and Potential Uses

Frank N. Jones[1] and W. (Marshall) Ming[2]

[1]Coatings Research Institute, Eastern Michigan University,
430 West Forest Avenue, Ypsilanti, MI 48197
[2]Eindhoven University of Technology, Den Dolech 2,
5612 AZ Eindhoven, The Netherlands

Convenient synthetic procedures for making nanoparticle latexes with particle diameters in the 12-20 nm range are reviewed. These procedures yield latexes that are more concentrated (20-30 wt. %) and have much less surfactant (8-15 wt. % of the monomer) than earlier procedures. Common acrylic and styrenic monomers, including cross-linkable monomers, can be used. Nanoparticle latexes have properties that are quite different from those of conventional latexes in at least four ways. (1) Their very large surface areas affect particle properties, (2) in some cases the polymer molecules are constrained within particles whose diameters are less than half the root-mean-square diameter of the same molecule in a larger particle, (3) if starved for surfactant, they can form gels in which the particle structure is largely intact, (4) in some cases the tacticity of the polymers is different from bulk polymers. These differences suggest unique applications.

Synthesis

Since 1980, when Stoffer and Bone first described microemulsion polymerization *(1)*, researchers worldwide have studied the process and sought commercial applications *(2)*. It is widely recognized that the method can be used to prepare latexes with particle diameters in the nanoparticle range (arbitrarily 5 – 50 nm). However, conventional microemulsion polymerization procedures require very large amounts of surfactant to produce particle diameters in the lower part of this range, and the products are produced at very low solids content. Sometimes there is twice as much surfactant as polymer in the products, and the polymer content is below 10 wt. %. These are major obstacles to practical application.

Synthetic procedures for low-surfactant, high-solids nanolatexes

In the late 1990s we set out to explore the limits of the method in hopes of minimizing these shortcomings. We found a modified microemulsion procedure that greatly improves the situation.

The new procedure is performed at mildly elevated temperatures under monomer starved conditions *(3,4)*. Typically, all water, surfactant, auxiliary surfactant (1-pentanol works well), and a small portion of the monomer are placed in a flask, purged with N_2, and warmed to 40 °C. All initiator is added followed by the major portion of the monomer, added continuously over a period of several hours. Gentle agitation is maintained throughout. The result is a stable, transparent dispersion of polymer nanoparticles in water. Particle diameters vary, but are often about 15 nm. The monomer/surfactant wt. ratio can be 7/1 to 12/1, and the polymer solids can be as high as 30 wt. %. In comparison, conventional microemulsion polymerization to give particle diameters in this range would probably require monomer/surfactant ratios of 1/2 and yield solids below 10 wt. %. A wide range of polymer and copolymer compositions can be made using many of the monomers commonly used in coatings.

Table I shows typical recipes for the process. In these cases, an anionic surfactant (sodium dodecyl sulfate) was used, and the initiator was ammonium persulfate catalyzed by N,N,N'N'-tetramethyl ethylene diamine. Non-ionic and cationic surfactants can also be used. Two conventional emulsion polymerizations, highlighted in gray, are shown for comparison. Monomers that yield both high and low T_g polymers (methyl methacrylate, MMA and butyl acrylate, BA) can be used along with butyl methacrylate (BMA) and methyl acrylate (MA). One of the MMA polymerizations (#970306) is highlighted because it stretches the limit of monomer/surfactant ratios that will yield a stable dispersion. The consequences of that are interesting, and will be described later.

Table I. Modified microemulsion polymerization, anionic surfactants

Monomer	Exp. No.	Initial Microemulsion Composition, g				Monomer added, g	Total Monomer content, g	Monomer/SDS ratio
		Monomer	SDS	1-pentanol	H_2O			
BMA	970610	0.5	1.4	0.2	88.4	9.5	10	7.1
BA	970506	0.5	1.4	0.2	83.4	14.5	15	10.7
	970318	2	1.4	0.2	84.4	12	14	10
MMA	970306	2	1.4	0.2	82.4	14	16	11.4
	970320 [a]	14	1.4	0.2	84.4	/	14	10
	970121	5	1	0.2	78.8	15	20	20
MA	970122	5	1	0.2	68.8	25	30	30
	970120 [a]	20	1	0.2	78.8	/	20	20

[a] Conventional emulsion polymerizations as comparisons.

Initiator: APS/TMEDA, 40 °C

Characterization of nanolatexes; particle growth during the process

Table II shows the particle diameters (measured on a L&N Microtrac Particle Analyzer, Model 9200) of the products from Table I. Diameter of one PMMA nanolatex was measured by transmission electron microscopy; results of the two methods were essentially identical (4). The right hand column shows the specific surface area (SSA), calculated from the measured number average diameters (D_n) of the particles assuming they are spheres. For comparison, PMMA and PMA latexes were made by conventional emulsion polymerization.

Figure 1 shows results of experiments in which samples were withdrawn and analyzed throughout the process using monomers with modest solubility in water. Clearly, new particles are formed throughout the process. They form, quickly grow to full size, and stop growing. This behavior is quite unlike conventional emulsion polymerization, where most of the particles form early on and grow throughout the process, with few new particles being formed in the middle and later stages. The quantities of ingredients in these experiments were the same as in Table I.

However, as shown in Figure 2, the situation is not as simple when monomers with very limited water solubility are used. Here the particles formed later in the process are larger, either because of changing reaction conditions or because some growth of the early particles occurs.

Hydrogel formation

What happens when we put in too much monomer? Something unusual. The PMMA nanolatex with a monomer/surfactant ratio of 11.4 (described in Table I) appeared perfectly stable at first, but after two weeks it turned into a transparent hydrogel, as pictured in Figure 3. PMMA nanolatexes with lower monomer/ surfactant ratios are stable indefinitely.

Table II. Particle size characterization of polymer nanolatexes

Polymer	Exp. No.	Dn (nm)	Dv/Dn	SSA (m2/g)
PBMA	970610	19.8	1.22	268.4
PBA	970506	16.5	1.34	287.6
	970318	13	1.26	404
PMMA	970306 [b]	15.3	1.47	329.8
	970320 [a]	49.4	1.09	115.1
	970121	14.9	1.47	325.6
PMA	970122	14.5	1.5	337.7
	970120 [a]	54.3	1.16	99.9

[a] Conventional emulsion polymerizations as comparisons
[b] Transparent **hydrogel** formed two weeks after polymerization

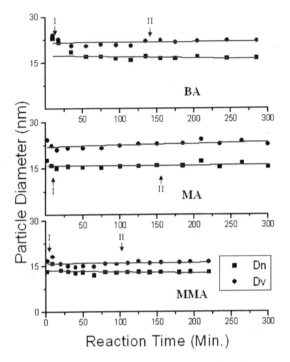

Figure 1. Particle size change versus reaction time during modified microemulsion polymerization of slightly water-soluble monomers (MMA, MA, and BA). "I" and "II" indicate the beginning and the end of monomer addition. (Reproduced from reference 3. Copyright 1998 American Chemical Society.)

Apparently this behavior occurs when the amount of monomer added gives polymer that just exceeds the amount of surfactant required to stabilize the dispersion. Extensive studies of the phenomenon *(5)* confirmed this hypothesis and revealed that gelation occurs with methacrylates (methyl, ethyl, and n-butyl) but not with acrylates (methyl, ethyl, n-butyl). Apparently only latexes with T_g's above room temperature can form hydrogels at room temperature. Other favorable factors for hydrogel formation include slightly soluble monomers (MMA, EMA) and relatively high initiator concentrations.

A cartoon illustrating hydrogel formation is shown in Figure 4 along with atomic force microscope images of dried PMMA nanolatexes before (a) and after (b) hydrogel formation. The images show that individual particles largely retain their identities, but the particles in Figure 4b appear to be slightly more connected. Once hydrogels form, the process is completely irreversible. This behavior suggests unique properties of nanolatexes and points the way to potential applications. A theory for why this behavior occurs will be advanced near the end of this paper.

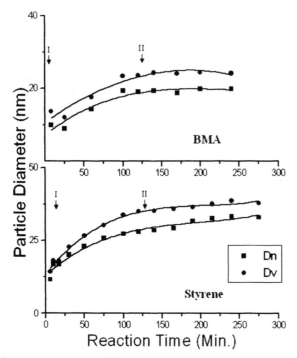

Figure 2. Particle size change versus reaction time during modified microemulsion polymerization of monomers (BMA and styrene) with low solubility in water. (Reproduced from reference 3. Copyright 1998 American Chemical Society.)

Figure 3. Typical nearly transparent PMMA hydrogels.
(Reproduced from reference 5. Copyright 1999 American Chemical Society.)

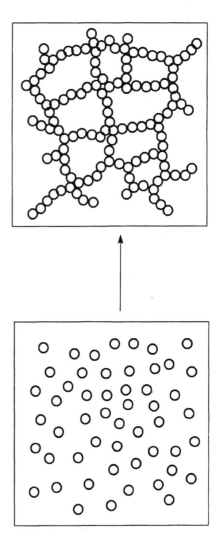

185

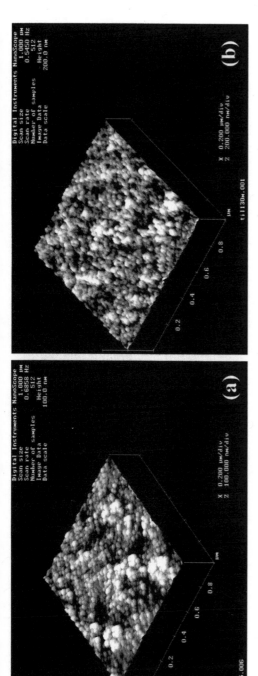

Figure 4. Schematic illustration of hydrogel formation (5) (upper) and AFM height images of PMMA nanolatexes before (a) and after (b) hydrogel formation. (Reproduced from reference 5. Copyright 1999 American Chemical Society.)

We compared (with several reasonable but unverified assumptions) the surface areas of the 15 nm PMMA particles with monomer/surfactant ratio 11.4 with the surface coverage expected of the SDS surfactant. The results led to a surprising conclusion. Apparently the critical monomer/surfactant ratio for hydrogel formation is reached when the surfactant level is only enough to cover about 25 % of the particles' surface. When coverage is even slightly higher, the nanolatexes are stable indefinitely.

Polymer Nanoparticle Properties Are Different

Pioneering research on properties of polymer nanoparticles made by microemulsion polymerization was done in Prof. Shoukuan Fu's laboratory at Fudan University in Shanghai (6). The early research used polystyrene nanolatexes made by conventional microemulsion polymerization. Several unusual properties were revealed. A summary of the results is included here to illustrate the ways in which the nanoparticles studied here may differ from conventionally made latexes of the same monomer composition.

The first thing to realize about these particles is that they may have very few polymer chains per particle. A particular nano-polystyrene had $D_n = 21.6$ nm and polymer $M_w = 980,000$. A simple calculation shows that there is room for only 3 polymer chains of this size per particle. Furthermore, as illustrated in Figure 5, the molecules must be packed into the particles, whose diameters are less than half the root-mean-square diameter of the same molecule if it were in bulk polystyrene. Presumably as a result of "unnaturally" confining the molecules, it was estimated that the density of polystyrene in similar particles is 9.5 % lower than in bulk polystyrene (7).

There are many other differences. Differential scanning calorimetry scans of nano-PS were different from that of conventional polystyrene, as shown in Figure 6. The top line shows the first scan of nano-PS, the middle line shows the second, and the bottom line shows conventional polystyrene. The first scan shows strong exotherms at 107 and 157 °C which disappeared in subsequent scans. The 107 °C exotherm can be attributed to surface energy released as the very small particles fuse. The 157 °C exotherm was "...ascribed to the relatively compact conformation of PS chains and the slightly ordered regions that consequently formed inside the particles..." (6). X-ray and infrared studies support this interpretation.

Recently, Prof. Fu's group reported extensive studies of nanoparticle latexes prepared by both the conventional (high surfactant) and modified (lower surfactant) microemulsion processes. Both PMMA (8) and polyethyl methacrylate (PEMA) (9) were investigated, with somewhat similar results. With PMMA, particle diameters were about 20 nm with both processes. M_w of the polymers in specific experiments were about 10^6 for the conventional method and 2.4×10^5 for the modified method. The average number of polymer chains per 20 nm particle was estimated at 2 for the conventional method and 11 for the

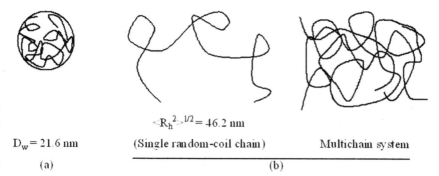

$\langle R_h^2 \rangle^{1/2} = 46.2$ nm

$D_w = 21.6$ nm (Single random-coil chain) Multichain system

(a) (b)

*Figure 5. Schematic illustration of conformations of (a) a polystyrene nanolatex
and (b) PS random coils with $M_w = 1 \times 10^6$. (Reproduced from reference 6.
Copyright 1996 American Chemical Society.)*

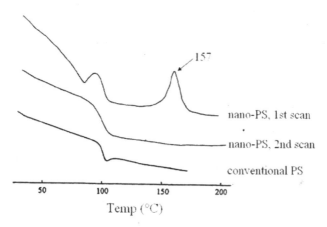

157

nano-PS, 1st scan

nano-PS, 2nd scan

conventional PS

Temp (°C)

*Figure 6. DSC curves of nano-PS (first and second scans) and a conventional PS.
(Reproduced from reference 6. Copyright 1996 American Chemical Society.)*

modified method. PMMA prepared by the modified method had remarkably low
polydispersity ($M_w/M_n = 1.4$), while the conventional method gave $M_w/M_n = 5.4$.

The polymer molecules in PMMA and PEMA nanoparticle latexes are
structurally different from those made by ordinary emulsion and bulk
polymerization. T_g of the nanoparticle PMMA from both processes was higher
than that of normal PMMA, generally around 125 °C. NMR study showed that
the nanoparticle latexes were rich in syndiotactic polymer (55 – 61 % for
PMMA and 58-64 % for PEMA), presumably explaining the T_g difference. It is
theorized that, in microemulsion polymerization, the tacticity of the growing
polymer chains is altered because much of the chain growth occurs at, or very
near, the polymer surface.

End Uses – What are Nanoparticle Latexes Good For?

It is clear from the above information that nanoparticle latexes are in many ways different from conventional latexes. How could one capitalize on the differences to make new and different materials and products? Efforts in this direction are underway.

Cross-linked and cross-linkable nanolatexes

It is possible to use the modified microemulsion polymerization method to make nanolatexes that are cross-linked within the particles and/or bear reactive groups that can cross-link later *(10)*. Monomer compositions of a few such materials are shown in Table III along with particle sizes and polymer contents. A control latex made by conventional emulsion polymerization is also shown. Nanolatexes can be made with ethylene glycol dimethacrylate (EGDMA), which causes cross-linking within the particles ("pre-coalescence cross-linking"). They can also be made with functional monomers, in this case acetylacetatoethyl methacrylate (AAEMA) and glycidyl methacrylate (GMA), which can be cross-linked by baking film cast from the nanolatexes ("post-coalescence cross-linking").

The primary reaction thought to be involved in post-coalescence cross-linking is shown in Figure 7. While there are possibilities of side reactions (*e.g.* GMA with with surfactant or initiator) and of premature cross-linking during polymerization at 40 °C, the data to be presented below in Tables IV and V suggests that such reactions are slow at temperatures below 85 °C. Thus, it is thought that most of the GMA/AAEMA cross-linking occurs during baking long after coalescence of the particles.

Table IV shows properties of coating films cast from nanolatex NL-1 on phosphate treated steel panels. The coatings had quite good properties when baked. The excellent resistance to rubbing with solvent (MEK) indicates a high degree of cross-linking within the films

Table V shows properties of films cast from the latexes described in Table III. NL-2 is capable of post-coalescence cross-linking but lacks internal cross-linking; its coating properties are good in some respects, but not as good as those of films cast from NL-1 (compare Table IV). NL-3 has internal cross-linking but lack capacity for external cross-linking; its film properties are poor.

Comparison of the films cast from nanolatex NL-1 (Table IV) with those cast from conventional latex CONV-L-1 (Table V) reveal remarkable contrasts. **These two latexes have identical monomer compositions, yet the properties of the films cast from NL-1 are substantially better than those of folms from CONV-L-1.** These latexes differ mainly in particle diameter (15 nm and 56 nm, respectively) and in surfactant content (NL-1 is higher). It is theorized that differences in particle diameter account for the property differences. The extra surfactant in NL-1 would be expected to degrade film properties, not improve

Table III. Functionalized polymer nanolatexes

Latex No.	Composition (wt%)	D_n (nm)	D_v/D_n	Polymer content (wt%)
NL-1	MMA, 36.1; BA, 46.3; GMA, 5.7; AAEMA, 8.6; EGDMA, 3.3	15.1	1.4	24.4
NL-2	MMA, 37.3; BA, 48.0; GMA, 5.7; AAEMA, 9.0	15.1	1.3	24.4
NL-3	MMA, 34.0; BA, 57.0	14.3	1.3	15.1
CONV-L1	MMA, 36.1; BA, 46.3; GMA, 5.7; AAEMA, 8.6; EGDMA, 3.3	56.3	1.04	24.4

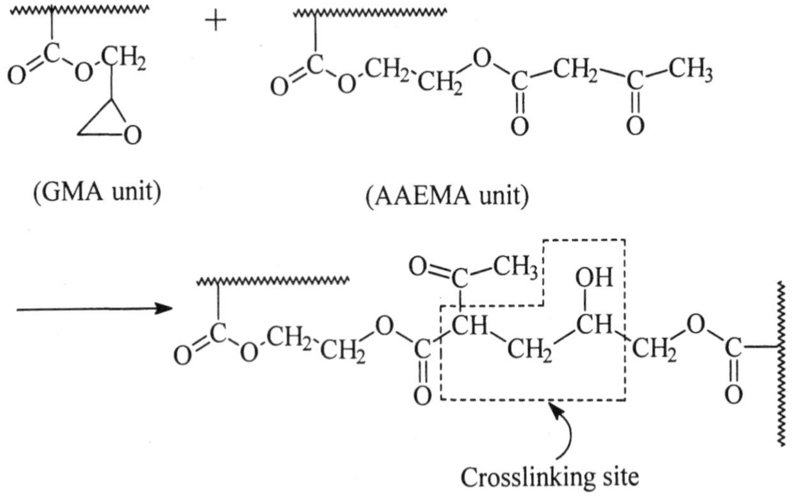

Figure 7. Crosslinking mechanism between GMA and AAEMA units in functionalized nanolatex films (10).

them.. Part of the differences in hardness, adhesion, and impact resistance can be attributed to different film thickness, but the differences are too large to attribute to film thickness alone. The observed property differences, especially in solvent resistance, suggest that the presence of pre-coalescence cross-linking in the 56 nm particles interferes with the ability of post-crosslinking to make films of sufficient uniformity for good properties. On the other hand, with 15 nm particles it appears that the combination of pre- and post-coalescence cross-linking is very effective. While general conclusions can not be drawn from one pair of experiments, **there is good reason to further investigate the possibility that nanoparticles can impart superior film properties in cross-linked latexes.**

Nanolatexes without conventional cross-linking

What about nanolatexes that are not functionalized? Their small size (15-25 nm *vs.* >50 nm for conventional latexes) alone results in a manifold increase in surface area and an increase in surface free energy in the system. Surface energy reduction is known to promote particle coalescence *(11)*. Furthermore, when molecular weights are high, molecules within the nanoparticles cannot extend to their usual root-mean-square dimensions because the particles are too small. It is theorized that this could create an additional, entropic, driving force favoring particle coalescence and rapid interpenetration. Such interpenetration would be driven by entropy increase as the constrained molecules extend to their preferred dimensions.

Table IV. Properties of films from NL-1, a nanolatex capable of both pre- and post-coalescence cross-linking

Property	Method	Unbaked film	Baked films 85 °C 1 h	Baked films 140 °C 30 min
Film thickness (mil)	ASTM D1186-87	0.9	1.0	0.9
MEK rub resistance	ASTM D3363-74	10	60	160
Pencil hardness	ASTM D4752-87	2B	2B	B
Impact resistance (in-lb), D/R	ASTM D2794-84	120/100	140/140	160/160
Adhesion	ASTM D3359-87	4B	2B	4B-5B
Tg (°C)	DSC	14.8	14.5	15.8

Table V. Properties of films cast from NL-2, NL-3, and CONV-L-1

	Film thickness (mil)	MEK rub resistance	Pencil hardness	Impact resistance (in-lb), D/R	Adhesion
NL-2 (no pre-coalescence cross-linking)					
Air dried	0.9	6	5B	120/80	4B
85 °C, 1 hour	0.9	25	2B	140/120	3B
140 °C, 30 min.	0.8	45	2B	160/160	3B
NL-3 (no post-coalescence cross-linking)					
Air dried	1.6	2	4B	20/20	1B
85 °C, 1 hour	1.8	4	4B	20/20	1B
140 °C, 30 min.	1.5	6	3B	20/20	1B
CONV-L-1 (conventional latex with with same composition as NL-1)					
Air dried	1.5	2	6B	40/20	1B
85 °C, 1 hour	1.8	6	6B	40/40	1B
140 °C, 30 min.	2.0	40	6B	60/40	1B

In view of these considerations, an explanation of the hydrogel formation described in Section II can be postulated. Association of marginally stabilized particles could be driven by reduction of surface free energy, a driving force that could occur with conventional latex particles but is magnified by the large surface area of nanoparticles. Once the particles are in contact, interdiffusion of molecules from different particles could be driven by an increase in entropy, perhaps facilitated by plasticization of the particles by water. The entropically driven process would knit the particles together irreversibly. It is unique to nanoparticles with few chains per particle. This postulate suggests applications in particle coalescence, adhesion, and perhaps drug delivery.

Acknowledgements

Support of this work by the National Science Foundation Industry/University Cooperative Research Center in Coatings at Eastern Michigan University is gratefully acknowledged.

References

1. Stoffer, J. O.; Bone, T. J. *Polym. Sci. Chem. Ed.* **1980**, *18*, 2641.
2. Kumar, P.; Mittal, K. L., Eds. *Handbook of Microemulsion Science and Technology,* Marcel Dekker, NY, NY, 1999.
3. Ming, W.; Jones, F. N.; Fu, S. *Macromol. Chem. Phys.* **1998**, *199*, 1075.
4. Ming, W.; Jones, F. N.; Fu, S. *Polymer Bull.* **1998**, *40*, 749.
5. Ming, W.; Zhao, Y.; Cui, J.; Fu, S.; Jones, F. N. *Macromolecules* **1999**, *32*, 528.
6. Fu, S. *et al. Macromolecules* **1996**, *29*, 7678.
7. Wu, C. *et al. Macromolecules* **1995**, *28*, 1592.
8. Jiang, W.; Yang, W.; Zeng, X.; Fu, S. *J. Polym. Sci. [A]: Polym. Chem.* **2004**, *42*, 733.
9. Tang, R.; Yang, W.; Wang, C.; Fu, S. *J. Macromol. Sci. [A]: Pure and Appl. Chem.* **2005**, *42*, 291.
10. Ming, W.; Zhao, Y.; Fu, S.; Jones, F. N. *Polym. Mater. Sci. Eng.* **1999**, *80*, 514.
11. Kan, C. S. *J. Coat. Technol.* **1999**, *[71]896*, 89.

Chapter 10

Preparation, Structure and Properties of Organic–Inorganic Nanocomposite Coatings

Shuxue Zhou, Limin Wu*, Bo You, and Guangxin Gu

Department of Materials Science, The Advanced Coatings Research Center of China Education Ministry, Fudan University, Shanghai 200433, People's Republic of China

Organic-inorganic(O/I) nanocomposite coatings have been developed very quickly in recent years and are gradually becoming one of the most active areas in coatings' research. This review classifies the preparation methods into three categories: embedding with nanoparticles, sol-gel process and self-assembly method. All these three routes are systematically discussed including the factors influencing the structure and properties of nanocomposite coatings. Some evaluations and suggestions are given for each method.

Organic-inorganic(O/I) nanocomposite coatings contain at least one phase with at least one dimensional size less than 100nm. Compared to the traditional coatings obtained from resins, micro-fillers and additives, O/I nancomposite coatings can more efficiently combine the advantages of rigidity of the inorganic phase and softness of the organic phase, or endow the coatings with new functionality resulting from nano-effects or synergistic effects. Therefore, more and more papers and patents involving with nanocomposite coatings have been published. At the very beginning, however, the concept of "nanocomposite coatings" was ever controversial, some people even doubted the value of nanomaterials used in coatings and the competitive power of nanocomposite coatings in market. Nowadays, more and more studies indicate that nanocomposite coatings can find potential uses in abrasion and scratch resistant coatings, corrosion resistant coatings, transparent coatings, superhardness coatings, functional coatings such as antibacterial coatings, UV-shielding coatings, conductive coatings, self-cleaning coatings, photocatalytic coatings and etc., some of them have even been commercialized. For example, BASF recently announced the use of scratch resistant automotive coatings containing nano ceramic particles (*1*). NEI (Nano Engineered Innovation) Corporation developed hard ceramic-like nanocomposite coating with trademark of NANOMYTE^{TM} for the plastic substrate such as poly(methyl methacrylate) (PMMA) (*2*). Obviously, nanocomposite coatings are gradually becoming one of the most quickly developing areas in coatings.

The related development of nanocomposite coatings was recently reviewed by Fernando (*3*) and Baer. et al. (*4*), respectively, but Fernando's paper was only a simple introduction of nanocomposite coatings while Baer et al.'s review mainly focused on inorganic nanostructure coatings. The objective of this review paper is focused on the preparation, structure and properties of O/I nanocomposite coatings and attempts to academically elucidate their relationships. Based on the literatures, we divide the preparation of O/I nanocomposite coatings into three methods: embedding with nanoparticles, sol-gel process, and self-assembly method. All the following discussions are carried out respectively based on these three methods. Meanwhile, some evaluations and suggestions are also given for each preparation method.

Nanocomposite Coatings Prepared Through Embedding with Nanoparticles

There are a lot of nanoparticles which could be used for improving the performance of coatings, different kind of nanoparticles may have various properties, even for the same kind of nanoparticles, the different source of nanoparticles (e.g., from powders or nanoparticles in sols), particle size, content and preparation method have also various influneces on the properties of coatings.

The Types of Nanoparticles and Their Purposes in Coatings

The frequently studied nanoparticles in coatings include nano-silica (nano-SiO_2), nano-tiantia (nano-TiO_2), nano-zinc oxide (nano-ZnO), nano- calcium carbonate (nano-$CaCO_3$), nano-alumina (nano-Al_2O_3), nano-zirconia (ZrO_2) and so on, which have been commercially available in the state of nanopowders or nanoparticle sols. Table I summarizes the physical properties and potential uses of these nanoparticles.

Table I. Summary of the physical properties and potential uses of commercially available nanoparticles.

Nanoparticles	Mohn's hardness	Refractive index	Potential uses in coatings
Nano-SiO_2	7	1.47	Scratch and abrasion resistance, UV-shielding, hardness
Nano-Al_2O_3	9	1.72	Scratch and abrasion resistance, UV-shielding, hardness
Nano-ZrO_2	7	2.17	Scratch and abrasion resistance, UV-shielding, hardness
Nano-ZnO	4.5	2.01	UV-shielding, fungus resistance, other functionality
Nano-TiO_2	6.0-6.5	2.70	UV-shielding, fungus resistance, other functionality
Nano-$CaCO_3$	3	1.49 and 1.66	Mechanical property, whitening

The inherent rigid nanoparticles such as nano-SiO_2, nano-Al_2O_3, nano-ZrO_2 particles and etc. are adopted to mainly improve the mechanical properties of traditional coatings. Among of them, nano-SiO_2 particles were the first nanoparticles produced. It was found that nano-SiO_2 particles could more efficiently increase the macro hardness, scratch resistance, elastic modulus of acrylic based polyurethane coatings than micro silica particles (5) or enhance the microhardness and abrasion resistance of polyester based polyurethane coatings (6). Zhang et al. (7) found that nanosilica could reduce the wear rate and frictional coefficient of the epoxy coatings even at low filler loading (<2vol%). Petrovicova et al. (8) found that improvements of up to 35% in scratch and 67% in wear resistance were obtained for coatings with nominal 15 vol % contents of hydrophobic nano-silica relative to unfilled nylon11 powder coatings. Fumed silica, one of nano-silica with primary particle size in nano scale range (10~20nm), is usually acted as the rheological additive in coatings. However, recent reports showed that it could also improve the abrasion and scratch resistance of the UV-curable coatings (9), the hardness and abrasion resistance of the acrylic or polyester based polyurethane coatings (*10,11*). Colloidal silica, another type of nano-silica particles, is also often used to improve the hardness, abrasion and scratch resistance of coatings. For example, Soloukhin et al. (*12*) found that nano colloidal silica particles could enhance the elastic modulus and hardness of acrylic-based UV-curable coatings on polycarbonate substrate. G. Chen et al. (*13*) indicated that nano colloidal silica could increase the storage modulus and abrasion resistance of acrylic base polyurethane coatings. Y. Chen et al. (*14*) presented increases in hardness, abrasion resistance and Tg for polyester-based polyurethane coats. Besides mechanical properties, heat resistance and barrier properties can also be improved by addition of nano-SiO_2 particles, which was observed in its application in nylon11 coatings (*8*) and acrylate radiation-curable coatings (*15-16*). Mennig et al. even found that nano- SiO_2 could provide both high scratch resistance and sufficient free volume for fast switching dyes of 3-glycidoxypropyltrimethoxysilane (GPTMS) pre-hydrolysate based coatings and thus developed a novel switching photochromic coatings for transparent plastics and glass (*17*). Nano-Al_2O_3 and nano-ZrO_2 can also improve the hardness, abrasion and scratch resistance of coatings and even better than nano-SiO_2 (*18*), but their refractive indexes (1.72 for nano-Al_2O_3 and 2.17 for nano-ZrO_2) are higher than that of nano-SiO_2 (1.46). Therefore, the coatings embedded with nano-Al_2O_3 and nano-ZrO_2 powder are usually semitransparent if these particles are not completely dispersed into nanoscale level (<100nm). However, the transparency of nano-Al_2O_3 containing composite coatings is easily attained in resins with high refractive index such as melamine-formaldehyde resins (*19*). Additionally, the relative price factor also influences the applications of nano-SiO_2, nano-Al_2O_3 and nano-ZrO_2 in coatings.

Nano-$CaCO_3$, the cheapest nanoparticles, can also enhance the hardness of organic coatings (*20*). Yan et al. (*21*) found that combining PVDF with nano-$CaCO_3$ would lead to super hydrophobic coatings (water contact angle>150°) while pure PVDF coatings had only 108°of water contact angle. However, the poor acid fastness of nano-$CaCO_3$ and the low transparency of nano-$CaCO_3$ - containing coatings considerably confine its application in coatings. Hitherto, nano-$CaCO_3$ is mainly used in polyvinyl chloride (PVC) anti-chipping coatings for automobile and paper coatings.

Nano-ZnO and nano-TiO_2 seem to have similar properties because of their semiconductive characteristic. Differing from nano-SiO_2, nano-Al_2O_3 and nano-ZrO_2, however, nano-ZnO and nano-TiO_2 are preferentially employed to improve the optical property or to offer some functional properties of the coatings. For example, Xiong et al. (*22*) found that nano-ZnO particles had UV-shielding property for acrylic latex coatings. Li et al. (*23*) indicated that nanosized ZnO particle could be acted as UV absorbent for the inorganic-organic hybrid coatings derived from tetraethoxyasilane (TEOS) and GPTMS on PMMA substrate. Xu et al. (*24*) found that tetrapod-like nano-particle ZnO had both electrostatic and antibacterial property and could be used to prepare multi-functional acrylic coatings. Nano-TiO_2, mainly in rutile, also has UV-shielding property (*25-27*), but it seems that nano-ZnO powder has higher UV-shielding property than nano-TiO_2 at the same level based on our study (28), while anatase nano-TiO_2 possesses outstanding photocatalytic property, which can be used to develop photocatalytic self-cleaning paints and anti-biological coatings (26). Other nanoparticles which can be used in coatings include: nano-Ag loaded silicate powder for antibacterial coatings, barium titanate ($BaTiO_3$) nanoparticles for ferroelectric coatings (16), nanoscale boehmite fillers for corrosion- and wear-resistant polyphenylenesulfide coatings (29) and nanocarbon tube and nanosized antimony-doped tin oxide (ATO) particles for conductive coatings (30).

The Preparation and Morphology of Nanocomposite Coatings Using Nanoparticles

Nanoparticles in nanopowder state usually aggregate due to their large specific surfaces. They can only work very efficiently at nano scale. Therefore, how to disperse nanoparticles in resins or coatings to their primary particles (nano scale) becomes an extremely tough problem that researchers and engineers are facing. In fact, some nanoparticles are relatively easily dispersed into polymer matrix depending upon the surface property of nanoparticles and the characters of the system (e.g., water-borne, solvent-base, hydrophilic or hydrophobic polymers, etc.), but in most cases, nanoparticles need surface treatment by using physical or chemical interaction according to the feature of continuous phase in resin or coating systems. Nevertheless, this treatment is usually combined with some extra powers such as grinding, ultrasonic treatment in order to acquire ideal dispersion. For example, Xiong et al. (22) successfully prepared nano-ZnO dispersions at primary size with high molecular weight block copolymer containing pigment affinic groups as dispersants through ball milling. Liu et al. (31) obtained nano-TiO_2 slurry at a mean particle size of 110nm using a modified amine with long aliphatic chains as the dispersant under the help of grinding. Fumed silica was dispersed into acrylic polyol resins by using vigorous stirring (10) and high pressure (more than 1 MPa) jet dispersion using at least one nozzle (32).

Nanoparticles can either be used into monomers through *in-situ* polymerization method to prepare nanocomposite resins, or directly applied into resins or coatings by mixing process. Figure 1 summarizes the possible routes for preparation of water-borne or solvent-based nanocomposite coatings from nanopowders.

For powder coatings, ball milling is usually adopted. For example, Petrovicova et al. (33,34) used ball milling and high-velocity oxy-fuel thermal spray processing to produce ceramic/polymer (silica/nylon) nanocomposite coatings. By optimizing spray parameters such as nozzle design, spray distance, oxygen-to-fuel ratio, powder feed position, and substrate cooling, dense coatings with relatively uniform particulate distribution could be achieved.

However, no matter how nanoparticles from nanopowders are treated, e.g., surface modification, grinding, ultrasonizing, or combination of these treatments, they are very difficult to fully disperse into nano scale level (<100nm), some of them, quite a few nanoparticles are still present in aggregates or agglomerates.

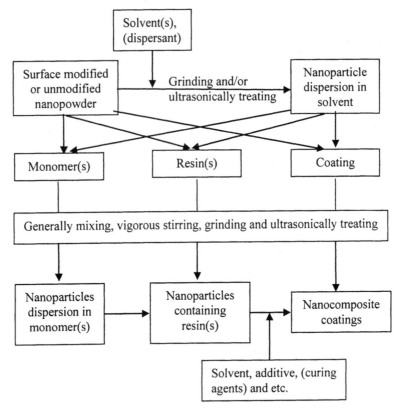

Figure 1. The possible routes for preparation of nanocomposite coatings from nanopowders

Figure 2 demonstrates the typical morphology of obtained nanocomposite coatings, which is composed of primary particle, the aggregates with size less than 100nm and the aggregates/agglomerates with size large than 100nm or even up to micrometer.

In comparison with nanopowders, the nanoparticles in sols are usually present in nano scale size, thus they can be directly employed into monomers, resins or coatings by simple mixing procedure. For example, Chen et al. (*35*) mixed colloidal silica with diol and diacid monomers and prepared polyester-based polyurethane/silica nanocomposite coatings via *in situ* polymerization. However, it should be noted that good compatibility between colloidal particles and both solvent and resins are necessary. Otherwise, aggregation can occur in the mixing and/or drying process (*36*). To work out this problem, surface modification of colloidal particle is also generally adopted. Figure 3 manifests the typical morphology of nanocomposite coatings from colloidal nanoparticles. Nearly all of the colloidal nanoparticles are evenly and individually dispersed in organic matrix.

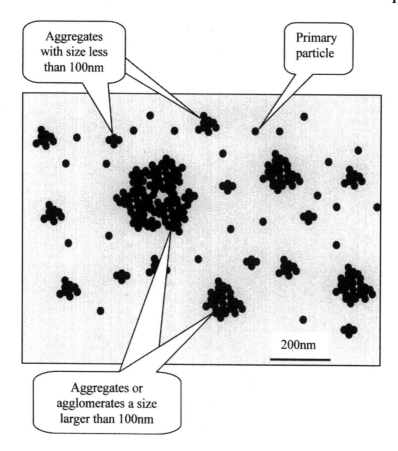

Figure 2. The schematic diagram of typical morphology of nanocomposite coatings from nanopowder

Based on the discussion above, it can be found that surface modification is usually necessary to improve the dispersibility of nanoparticles and their compatibility with polymer matrix and/or solvent no matter the nanoparticles are from powders or colloidal particles. The surface modification can be conducted physically through adsorption of surfactant and deposition of inorganic substance, or chemically via reaction with coupling agent, alcohol, or macromolecule. Among the chemical modification, silane coupling agents (SCA) are the most frequently used. The end-group of SCA molecule, alkoxyl group, can react with the hydroxyl groups of nanoparticles, the organic chain without/with end-groups such as vinyl, epoxide, amine, isocynate and mercaptan, can interact with polymers, or with nonfunctional organic chains which can be compatible with organic phase. For instance, for favoring the incorporation of nanoparticles into the polyacrylate matrix, the surfaces of nano-sized silica and

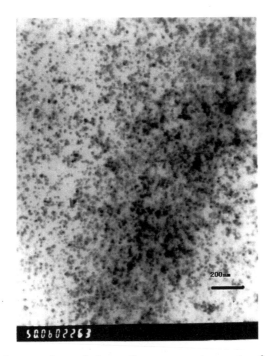

Figure 3. The typical morphology of nanocomposite coating from colloidal particles. (Reproduced with permission from reference 35. Copyright 2003 John Wiley & Sons)

alumina particles were chemically modified by reaction with methacroyloxy (propyl)-trimethoxysilane (MAPTMS) (*37*). The formation of covalent Si(Al)-O -Si-C bonds between MAPTMS and silica or alumina was demonstrated by means of FTIR and ^{29}Si NMR spectroscopy. These modified nanoparticles then formed covalent bonds with acrylic resin under UV radiation, efficiently improving viscoelastic properties. In addition, GPTMS was also adopted to modify nano-SiO$_2$ particles to investigate the effect of surface property of nanoparticles on the Young's modulus and scratch resistance of polysiloxane coatings (*38*).

The Factors Influencing the Properties of Nanocomposite Coatings

The properties of nanocomposite coatings mainly depend on the composition of organic matrix and inorganic nanophase, and also the dispersion of nanoparticles and interfacial interaction between polymer matrix and nanoparticle. Here we would just focus on the effects of the properties, surface modification and content of nanoparticles on coating properties.

Effect of the Properties of Nanoparticle

The properties of nanoparticle such as types, sizes, crystal structures and so on have obvious impact on the properties of nanocomposite coatings. Our series studies showed that nano-SiO_2 from nanopowders could more remarkably increase the hardness, scratch resistance, abrasion resistance and UV shielding properties of coatings than colloidal silica from sol-gel process, but the latter seemed to keep reasonable viscosity and gloss of coatings in comparison with the former (*5,6,11,13*). Bauer et al. (*18*) found that the abrasion resistance of UV-curable nanocomposite coatings was related to the types of nanoparticles. Those nanoparticles with high Mohn's hardness, e.g., alumina, silica (quartz), and zirconia, improved the abrasion resistance of UV-curable coatings more efficiently than the nanoparticles with low Mohn's hardness, e.g., titania. pyrogenic silica, though with thickening effect, showed much greater improvement in the surface mechanical properties such as haze and diamond microscratch hardness than colloidal silica. The primary particle size not only affects the optical property but also the mechanical property of nanocomposite coatings since it is one of the most important parameters influencing the interfacial interaction. Chen et al. (*14,36*) investigated the influence of the colloidal silica particle size on the mechanical properties of acrylic or polyester based polyurethane coatings and found that the abrasion resistance, Tg and hardness first increased then decreased with increasing colloidal silica particle size in the range of 10~200nm (*14*), the optimal particle sizes of colloidal silica were within 30~70nm range for effectively improving the mechanical properties of acrylic-based or polyester-based coatings. This was probably because much smaller colloidal silica particles easily caused some aggregation, which contrarily decreased the interfacial interaction, while much larger silica particles weakened the interfacial interaction. Xiong et al. (*22*) found acrylic latex paint embedded with 60 nm ZnO particles has higher Tg and better UV-shielding property than those of 100nm ZnO. The crystal structure of nanoparticles also strongly influences the properties of nanoparticles and thus their embedded nanocomposite coatings. As mentioned above, anatase nano-TiO_2 has very strong photoactivity which can be used for anti-bacterial or self-cleaning coatings (*26*) while rutile nano-TiO_2 has relatively high UV-shielding property which is mainly applied for the preparation of high weathering resistant coatings (*25*).

Effect of Surface Modification of Nanoparticles

Surface properties of nanoparticles have a very strong impact on the interfacial interaction between organic matrix and inorganic phase, thus influences the properties of coatings. More and more research indicates that surface modification of nanoparticle is favorable for enhancing coatings performance since surface modification avails increasing interfacial physical or chemical interaction between nanoparticles and organic matrix, and obtaining better dispersion of nanoparticles in organic matrix. Zhang et al. (*7*) found that embedding nano-silica into epoxy could reduce wear rate and frictional coefficient of the matrix at low filler loading, more obvious improvement of the tribological properties was obtained by incorporating polyacrylamide (PAM) grafted nanoparticles. Schadler et al. (33) found that silica particles with a hydrophobic (methylated) surface resulted in better mechanical properties of nylon coating than those with a hydrophilic (hydroxylated) surface. Bauer et al. (18) indicated the benefit of surface silylation of pyrogenic inorganic filler in enhancing the respective surface mechanical properties of radiation cured coatings. Chen et al. (13) found that higher abrasion resistance of acrylic based polyurethane coatings was obtained with SCA modified silica compared with the unmodified one. However, different silylation surface chemistry of silica did not substantially influence the scratch and abrasion resistance of radiation cured coatings with 15wt% silica load as well as acrylic based polyurethane/colloidal silica nanocomposite coatings (13,39). This is probably attributed to the formation of ladder-like polysiloxane structures on the surfaces of the filler particles during organophilation of nanoparticles.

Effect of the Nanofiller Volume Fraction

Generally, the mechanical properties of coatings, such as abrasion and scratch resistance, hardness and elastic modulus, increases with increasing nanoparticle content. But different research has reported various nanoparticle content required to acquire the desirable properties of coatings. Some indicated only small amount of nanofillers could obviously improve the performances. For example, Zhang et al. (7) found that only 2 vol% of PAM modified nano-silica could efficiently decrease wear rate and friction coefficient of epoxy coatings. Zhou et al. (5) found that only 1wt% (about 0.5 vol%) nanosilica embedded could enhance around 20% of microhardness for acrylic based polyurethane coatings. But other researchers found that high filling content should be used to obviously improve the coating properties. For example, improvements of up to 35% in scratch and 67% in wear resistance were obtained for nylon coatings with nominal 15 vol % contents of hydrophobic silica (8), and 35wt% (about 23 vol%) colloidal nanosilica was used for substantial improvement of the scratch resistance of irradiation curable coatings (18). Theoretically, if the cost of nanocomposite coatings is not considered, higher nanofiller content generally favors enhancing coating properties, but more nanoparticle content unavoidably increase the viscosity of resins or coatings (mainly solvent-based), thus much higher VOC is needed for application. In addition, too much nanofillers could possibly impair the toughness of nanocomposite coatings.

The change of the structure and morphology of organic matrix caused by nanofillers may be another reason responsible for the change of the coating properties. For example, hydrophobic nanosilica or carbon black increased the crystallinity of nylon powder coating and thus increased the scratch resistance, wear resistance and storage modulus (8,35).

Some other factors such as preparation method, coating composition and etc., also have very important influence on the performances of nanocomposite coatings. Figure 4 summarizes some main factors affecting the properties of nanocomposite coatings. Obviously, it is very tough to prognosticate the performances of nanocomposite coatings since so many parameters work on the structure and properties of coatings.

Clay-containing Nanocomposite Coatings

Unlike the prepared nanoparticles, clay is the specific nanofiller whose nanostructure is *in situ* generated during preparation of nanocomposite coatings. Clay can also improve the hardness, scratch resistance, viscoelastic properties of coatings and provide the coating with corrosion resistance and barrier property. Yeh et al. (40-43) consistently reported the enhancement of corrosion resistance and barrier property of H_2O and O_2 by clay for different polymer matrix such as polyaniline (40), poly(o-ethoxyaniline) (41), PMMA (42), poly(styrene-co-acrylonitrile) (43), and etc. They prepared polymer-clay nanocomposites (PCNs) via *in-situ* polymerization with quaternary alkylphosphonium salt or quaternary alkylammonium salt as an intercalating agent or via solution dispersion technique (44,45). All PCNs had better corrosion resistance and barrier property than their corresponding pure polymer matrix at low loading (0.5~3%), which resulted from dispersing silicate nanolayers of clay in the polymer matrix to increase the tortousity of the diffusion pathway of oxygen and water. Meanwhile, higher Tg and thermal properties were also observed for most of the prepared PCNs. Schemper et al. (46) incorporated synthetic clay into UV-curable coatings using Jeffamines as polymer/clay compatibilizers. Earlier onset of auto- acceleration, higher polymerization rate and conversion were achieved, and hardness and scratch resistance increased for relative soft coatings (e.g., methyl α-hydroxymethylacrylate crosslinked with 1,6-hexanediol diacrylate), but did not vary for inherent hard coatings (e.g., methyl α-hydroxy-methylacrylate crosslinked with 3-(acryloyloxy)-2-hydroxypropyl methacrylate (AHM)). Chen et al. (47) investigated the influence of organoclay on the corrosion resistance of epoxy aircraft coatings. Incremental corrosion resistance was not observed because of the inherent high corrosion resistance of the studied epoxy coatings although nanocomposite coating with exfoliated structure was obtained for the fast curing rate of epoxy coatings. Therefore, it seems that clay is helpless for those organic coatings with originally excellent mechanical and/or corrosion resistance properties. Besides the studies on the effect of clay on mechanical and corrosion resistant properties of organic coatings, some clay-containing nanocomposite coatings for special application were also reported. For example, Majumdar et al. (48) developed an aqueous transparent nanocomposite coatings consisting of Laponite (a commercial synthetic clay)/PVP/PEO (weight ratio 35:15:50) for application as a fast drying,

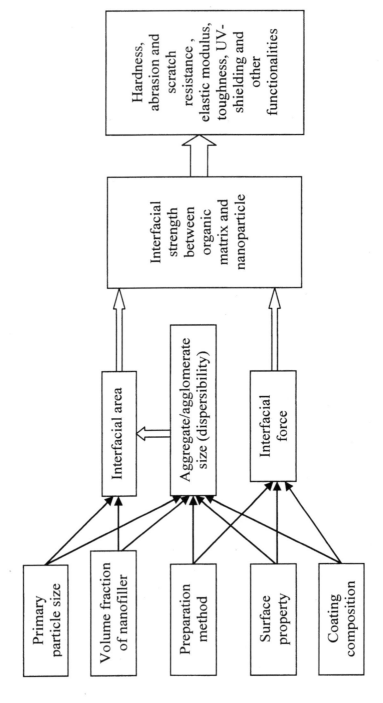

Figure 4. The hierarchy relationship of different factors influencing the properties of nanocomposite coatings

high quality, image-receiving layer for inkjet printing on a variety of substrates such as polyester, PE-coated or PP-laminated photo paper, GP paper and so on. Ranade et al. (49) prepared a mixed exfoliated and intercalated polyamide-imide nanocomposite with montmorillonite for application as magnetic wire coatings with reduced specific heat and improved Vicker hardness. Since the refractive index of clay (bentonite clay = 1.54) closely matches those of most organic coatings, and at most cases, the clay loading in coatings is very small (usually less than 5%), transparent nanocomposite coatings can be easily obtained from clay, and clay is very cheap. Therefore, clay-containing coatings may be worthy studying and developing especially for the improvement of inherent poor quality coatings.

Nanocomposite Coatings via Sol-gel Process

Sol-gel process is the earliest developed method to prepare nanocomposite (hybrid) coatings. Initially, sol-gel process was mainly used to develop high performance inorganic coatings whose thickness is usually very thin (in several hundreds nanometer scale) and the coatings with desirable properties can be formed only through baking at high temperature (50). However, the high baking temperature of inorganic coating restricts its application on some relatively soft substrates such as plastics or immobile articles, thus organic-inorganic hybrid coatings have been extensively studied based on sol-gel process in recent years. Compared with pure inorganic coatings, the properties of the hybrid coatings are easily controlled in a wide range by tailoring the parameters of sol-gel process. The hybrid coatings obtained from sol-gel process are usually transparent and generally acted as the abrasion and scratch resistant coatings for plastics (51), glass (52,53) and metals, optical coatings (54), flame retarding coatings (55), or the corrosion resistant coatings for metal substrates such as steel (56), aluminium (57) and magnesium. Figure 5 schematically compares the preparation methods of different sol-gel derived coatings. It can be clearly seen that the preparation of the hybrid coatings is more complicated than that of the ceramic coatings.

The Preparation of Sol-gel Derived Nanocomposite Coatings

Actually, the sol-gel process only provides the inorganic phase for nanocomposite coatings via the hydrolysis and condensation of silicone/metal alkoxide precursors. The precursors mainly include alkoxysilane, titanium alkoxide and aluminium isopropyl. Other metal alkoxides are seldom reported in the preparation of nanocomposite coatings because of their high reactivity in sol-gel process and limited availablity. Among all these precursors, the sol-gel process of alkoxysilane is relatively mild and easily controlled. Therefore, most of the sol-gel derived nanocomposite coatings are silica-based hybrid coatings.

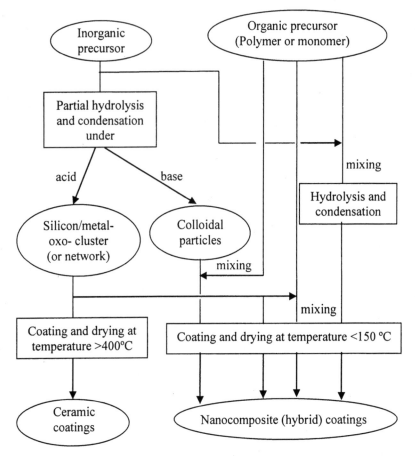

Figure 5. The simplified process of sol-gel method and the possible sol-gel derived coatings

Silica-based Hybrid Coatings

The commercially available alkoxysilanes include MAPTMS, GPTMS, vinyl triethoxysilane (VTES), methyl triethoxysilane (MTES), aminopropyltriethoxy- silane (APTES) and TEOS. Figure 6 demonstrates their chemical structures.

For the silica-based hybrid coatings, the inorganic phase can be obtained from three routes. One is the hydrolysis and condensation of trialkoxysilane. The sol-gel process of trialkoxysilane leads to organic-inorganic hybrid polyhedral oligomeric silsesquioxnes (POSS) which can act as nano filler or monomers (for functional POSS) for polymers (*58*) or individually form the coatings, however, the individual film-formability becomes poor when the organic substituted group enlarges because of steric hindered effect on the condensation of alkoxylsilane (*59*). Trialkoxylsilane precursors can also be firstly mixed with oligomer/ polymer, e.g., GPTMS in epoxidized plant oils (*60*), and then undergo hydrolysis and condensation with acid catalyst to prepare hybrid coatings. The second source of silica phase is from the hydrolysis and co-condensation of TEOS and/or tetramethoxysilane (TMOS) with one or more trialkoxysilanes. Since the pre-hydrolyzed product of TEOS and/or TMOS is incompatible with organic phase in most cases, the trialkoxysilane precursor should be used to enhance the interfacial interaction between organic and inorganic phases. Compared with the individual sol-gel process of trialkoxylsilane, the existence of TEOS or TMOS facilitates the formation of inorganic phase with large size and high compactness. Besides the above two sources, the sol-gel product of trialkoxylsilane or trialkoxylsilane/TEOS combining with some inorganic nanoparticles such as nano-boehmite (AlOOH) (*61*), nano-silica (*57,62*), nano-TiO$_2$ (*63*), nano-Al$_2$O$_3$ (*64*) nano-ZrO$_2$ (*64*) also consist the inorganic phase together. The incorporation of inorganic nanoparticles in sol-gel derived hybrid coatings will make the coatings denser and offer higher scratch resistance or better corrosion resistance.

The organic component in the preparation of hybrid coatings may come from three routes: (1) the organic groups provided by organotrialkoysilane or

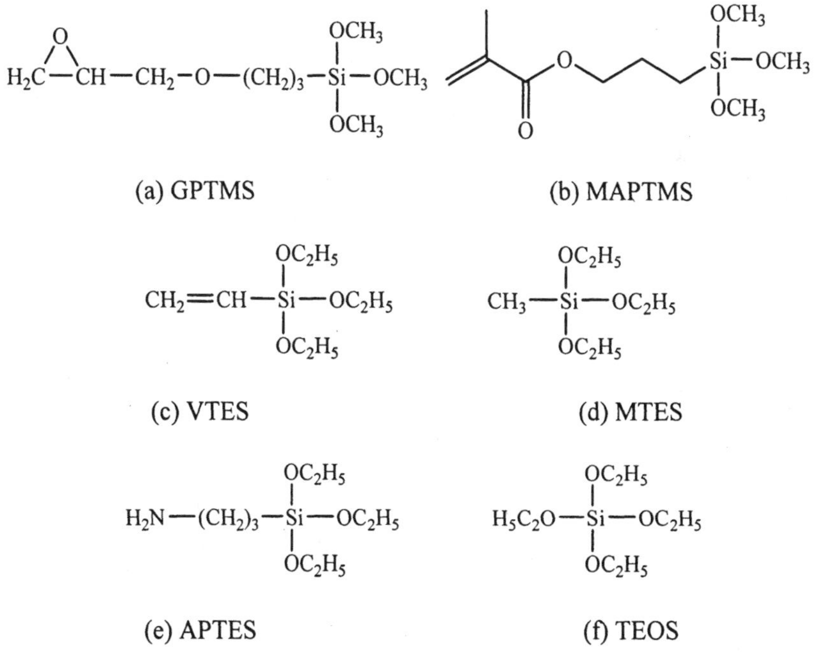

(a) GPTMS

(b) MAPTMS

(c) VTES

(d) MTES

(e) APTES

(f) TEOS

Figure 6. The chemical structure of common alkoxysilane

organic crosslinking agents such as DETA (*65*), Jeffamine (*62*), (2) organic monomers and (3) polymers or oligomers. For the second route, monomers are usually copolymerized with vinyl-containing trialkoxysilane (e.g., MAPTMS, VTES.) at first and then undergo a sol-gel process. For example, Harreld et al. (*66*) first synthesized the copolymers of MAPTMS and methyl methacrylate (MMA) with various composition and molecular weight for transparent hybrid materials. The reported hybrid coatings with polymer or oligomer as the starting material are polysiloxane/silicon-oxo-clusters protective space coatings for vehicles in low earth and geosynchronous orbit (*67*), biodegradable plant oil-silica hard hybrid coatings (*60*), acrylate end-capped polyester or polyurethane oligomeric resin-based UV curable hard hybrid coating for polycarbonate (*68*).

Titania-based Hybrid Coatings

Compared with silica precursor, the types of precursor for preparing the titania-based hybrid coatings are very limited. The commercially available precursors are tetrapropylorthotitanate (TPOT) (*54*), titanium n-butoxide(TBT) (*69*). Because of the high reactivity of titania precursors, chelating agents, e.g., acetylacetone, acetic acid, are usually necessary in the sol-gel process. Most of the reported coatings derived from titania precursor are actually ceramic coatings used as high refractive index coatings for optical apparatus, photocatalytic coatings (*70*) for self-cleaning purpose and water treatment (*71*). In recent years, however, titania based hybrid coatings have also been developed. For examples, Nguyen et al. (*72,73*) prepared titania-poly(MMA-co-butyl methacrylate-co- methacrylic acid) hybrid coatings on mild steel for corrosion resistant purpose. Kim et al. (*74*) developed a sol-gel titania-poly(dimethylsiloxane) (TiO_2-PDMS) coating as a sorbent in capillary microextraction with excellent pH stability and enhanced extraction capability over the commercial gas chromatograph coatings. Yeh et al. (*75*) produced PMMA-titania hybrid coatings using hydroxyethyl methacrylate (HEMA) as coupling agent and TBT as titania precursor. Xiong et al. (*69,76*) prepared acrylic-titania hybrids using carboxyl-functionalized or trialkoxylsilane-capped acrylic resin and pre-hydrolyzed titania sol. Chen et al. (*77*) also employed trialkoxysilane-capped PMMA to prepare titania-based hybrid optical coatings. Yuwono et al. (*78*) developed nanocrystalling TiO_2- PMMA hybrid for nonlinear optical coatings using titanium isoproproxide as the starting material for nanocrystalline titania, together with MMA and MAPTMS. Sarwar et al. (*79*) prepared titania-aramid hybrid materials through aminophenyl-trimethoxysilane end-capped aramid chains and TPOT. Furthermore, titania precursor and silica precursor can also be combined to create the inorganic phase (*80*).

Other metal alkoxide derived hybrid coatings are seldom reported because of the difficult control of the hydrolysis and condensation reaction. Chen et al. (*62*) consistently reported that the preparation of silica-based hybrid coating containing colloidal nano-Al_2O_3 particles via the sol-gel process of aluminum isopropoxide. Roux et al. (*81*) developed a new conductive zirconia based hybrid coating for ITO electrodes through the hydrolysis-condensation of zirconium tetrapropoxide $(Zr(OPr)_4)$ in presence of carboxylic acid or a β-diketone group functionalized pyrrole and electropolymerization.

Functional nanoparticles can also be incorporated into the hybrid coatings to develop the hybrid functional coatings. For example, Spirkova et al. (*82*) prepared conductive hybrid organic-inorganic coatings using GPTMS, amino-terminated poly(oxypropylene) (Jeffamine D-230) and colloidal polyaniline nanoparticles. Li et al. (*23*) prepared UV-shielding inorganic-organic nanocomposite coatings using TEOS, GPTMS and nano-ZnO particles.

The Structure and Properties of Sol-gel Based Hybrid Coatings

The structure of inorganic phase from sol-gel process strongly depends on the types of catalyst, pH values, water/precursor ratios, the compositions of precursor as well as baking temperatures of coatings. The sol-gel process of silicon/metal alkoxide under acid catalytic condition will form a structure totally different from base-catalyst, just as described in Figure 5. Under base catalytic condition, the condensation reaction of silicon/metal alkoxide is faster than its hydrolysis reaction, resulting in dense colloidal particle structure, and thus heterogeneous hybrid coatings can be obtained. Contrarily, under the acid catalyst, the condensation reaction is slower than the hydrolysis, which avails forming silicon/metal-oxo-cluster (namely network) structure, and thus transparent homogenous hybrid coatings can be obtained. The silicon/metal-oxo-clusters are open and mass fractal. More polymeric chains can be entrapped by the inorganic network compared with nanoparticle, which can obviously increase the interaction between organic and inorganic phases. For an instance, Deffar et al. (*83*) compared the influences of titanium-oxo-clusters and TiO_2 nanoparticles on the properties of hybrid coatings, and found that the titanium-oxo-clusters-based coats had higher tensile modulus, storage modulus, and Tg than the TiO_2 particles-borne coats.

It is certain that the chemical structure of organic phase influences the properties of the resulted hybrid coatings very much. Generally, crosslinking organic phase is preferred for enhancing the elastic modulus and hardness of hybrid coatings, but too much crosslinking maybe decreases the toughness. For example, Robertson et al. (*65*) found that the increment of DETA content increased both elastic modulus and hardness for the coatings derived from GPTMS/DETA. Spirkova et al. (*62*) studied the effects of the types of Jeffamine and the ratios of [NH]/[epoxy] on the Tg, tensile behavior, storage modulus, toughness of the GPTMS/GMDES/SiO$_2$/Jaffamine coatings. Hard coatings were obtained from GPTMS/Jaffamine T403 (poly(oxypropylene)s end-capped with three primary amino groups), but using toughness as the criteria, the coating formulation from GTMS/Jaffamine D230 (poly(oxypropylene)s end-capped with two primary amino groups) at the [NH]/[epoxy] ratio of approximately 1.5 was the preferred choice.

Interfacial interaction between organic phase and inorganic phase also greatly influences the properties of hybrid coatings. The chemical bonds between organic and inorganic phase, which commonly form using trialkoxysilane, HEMA, acrylic acid or other functional monomers containing organic phase with inorganic phase (69,75,76), can avoid phase separation, increasing the properties of hybrid coatings, such as abrasion and scratch resistance, hardness and Young's modulus.

Of course, different heat treatment can also have some impact on the hybrid coatings since relatively high temperatures make for further interaction between organic phase and inorganic phase or inorganic phase itself.

Nanocomposite Coatings by Self-assembly Method

Self-assembly method is a special method for fabricating O/I nano-composite coatings with excellent properties. In the past decade, three main self-assembly methods were developed, namely evaporation induced self-assembly (EISA) process, self-assembled nanophase particle (SNAP) coatings process and electrostatical self-assembly (ESSA) process.

Evaporation Induced Self-assembly (EISA) Process

EISA process was proposed and mainly studied by Sellinger et al. (84). In this process, a solution of silica, surfactant, organic monomers, and a mixture of solvents such as water and ethanol is initially formulated. The substrate is dip-coated with this solution. Micelles with partitioning of the organic constituents into the micellar interior and silica into the micellar exterior are formed by the evaporation of solvent during the drying process. Subsequent self-assembly of the silica–surfactant–monomer micellar species into lyotropic mesophases simultaneously organizes the organic and inorganic precursors into the desired hexagonal, cubic or lamellar mesoscopic form. Polymerization fixes this structure, completing the nanocomposite assembly process. A nanocomposite coating that mimic nacre was successfully prepared with this process using dodecylmethacrylate and hexanedioldimethacrylate as organic monomer and silica as inorganic phase (84). 1GPa of hardness, which was the indentation hardness measured for rather dense sol–gel silica films, could be achieved for even the coatings containing 50% polymer. Lu et al. (85) successfully prepared mesoscopically ordered chromatic polydiacetylene/silica nanocomposites via EISA method using oligoethylene glycol functionalized diacetylenic (DA-EOn) as polymerizable surfactants. The nanocomposite exhibited unusual chromatic changes in response to thermal, mechanical and chemical stimuli and meanwhile provided sufficient mechanical integrity to enable integration into devices and microsystems. Mesostructured conjugated poly(2,5-thienylene ethynylene)/silica nanocomposites were also prepared by EISA method (*86*). It was usually used as the coatings in light emitting diodes, information storage devices and optical signal processors. It seems that EISA is an ideal method to create mesoscopically ordered nanocomposite films (*87*). But it is regreted that the preparation condition of EISA process was very rigorous. The types of surfactant, the constitutes of precursor solution, the drying process and so forth all greatly affect the structure of coatings or films. In addition, EISA process is hitherto only applied on the substrates of silicon wafer and glass which possess very smooth surface. Thus these shortcomings of EISA process will severely cumber its practical application. A simple and feasible EISA process needs to be further developed for preparation of high performance coatings.

Self-assembly Nanophase Particle (SNAP) Coating Process

The SNAP coating process consists of three stages: (1) sol–gel processing; (2) SNAP solution mixing; (3) SNAP coating application and cure. The 'sol–gel processing' stage involves hydrolysis and condensation reactions of GPTMS/TMOS and is controlled by the solution pH and water content. In the 'SNAP solution mixing' stage, crosslinking agents and additives are added to the solution, which is then applied to a substrate by dip-coating to form the SNAP coating (*88*). The crosslinking agent, DETA, is also employed to improve the barrier properties and film forming ability of SANP coatings. However, amino-silanes offer more significant improvement in coating performance than DETA (*89*). The ordered, homogeneous and nanosized siloxane macromolecule nanostructure of SNAP coatings were confirmed by both molecular simulation (*90*) and the experimental data from XPS, X-ray diffraction (XRD), time-of-flight secondary ion mass spectrometry (TOF-SIMS) and atomic force microscopy (AFM) characterizations combined with a charge referencing method (*88,91*). SNAP coatings could be used as potential replacement for chromate-based surface treatments on aircraft aluminum alloys (*92,93*). Electrochemical impedance spectroscopy (EIS) tests demonstrated that SNAP coatings were defect-free, durable films with good corrosion protection. If SNAP coating was further combined with organic inhibitors such as aminopiperidine (APP), aminopiperazine (APZ), better corrosion protection of the coating would be achieved (*93*).

The self-assembling behavior (self-organized and long range ordered) of siloxane cagelike clusters was also observed in the epoxide-based products made from functionalized organosilica building blocks(GPTMS), functionalized oligo(oxypropylene)-diamine and/or -triamine, and colloidal silica nanoparticles under certain reaction conditions (*94*).

Electrostatical Self-assembly (ESSA) Process

In this process, polyelectrolyte is first assembled onto substrates. Then, the nanoparticles with opposite charge are electrostatically deposited on the polyelectrolyte surface. ESSA process has already been widely applied to fabricate core-shell materials, hollow capsules and hollow spheres (95,96), but seldom on the preparation of nanocomposite coatings.

Rosidian et al. (97) prepared superhard ZrO_2/polymer nanocomposite thin films through ionic self-assembly method. Silicon wafer or quartz substrate was first treated with poly(allylamine hydrochloride) (PAH) and a molecular dye (PS119) , and then colloidal ZrO_2 with poly(sodium-4-styrenesulfonate) (PSS) were sequentially deposited on the surface of PAH/PS119 via layer-by-layer process. The Vickers microhardness of as-deposited coatings at room temperature was better than that of ZrO_2 coatings sputter-deposited at 80□. If it was heat-treated, denser and higher hardness of the coatings would be further achieved. ESSA process was also studied by Prem et al. (98) to prepare nanocomposite coating with low friction. They first assembled poly(allylamine hydrochloride) (PAH)/poly(acrylic acid) (PAA) multilayer onto glass, silicon wafer, and stainless steel substrates, and then silver nanoparticles or multi-wall carbon nanotubes (MWNT) were electrostatically adsorbed via layer-by-layer process. It was found the coatings of PAH/PAA/silver nanoparticles had low friction at low normal load while the coatings of $((PAH/PAA)_1/ (PAH/MWNT)_5)_x$ demonstrated low friction and wear at all levels of stress. Zhai et al. (99) further found that the PAH/PAA multilayer was a honeycomb-like texture with surface pore as large as 10μm after too low pH treatments. They deposited 50 nm SiO_2 nanoparticles onto the surface of PAH/PAA multilayer via alternating dipping of the substrates into an aqueous suspension of the negatively charged nanoparticles and an aqueous PAH solution (about 20 cycles of absorption of SiO_2/PAH). The surfaces were further modified by chemical vapor deposition (CVD) of (tridecafluoro-1,1,2,2-tetrahydrooctyl)-1-trichloro-silane (semifluorinated silane). The final resulted stable superhydrophobic coatings that mimic lotus effect had both advancing and receding contact angles of 172°.

The above layer-by-layer electrostatical assembly process can easily control the thickness of the coatings by tuning the numbers of repeating self-assembly cycle of polyelectrolyte and colloidal particles. However, the process is time-consuming. To attain the thickness on micron level, tens or even hundreds of absorption cycle should be conducted. Hattori (100) simplified the electrostatical self-assembly process. He developed two-step assembly technique for preparation of polymer-particle composite films. Polyelectrolyte such as poly(diallyldimethyl-ammonium chloride) (PDDA) was spin-coated firstly on glass substrate and then colloidal silica (SiO_2) or titania (TiO_2) particles were adsorbed on the surface of polyelectrolyte by dipping in the particles suspension. Porous or even colloidal crystal containing composite films with thickness of several microns or even up to 37.9 microns were obtained finally. Moreover, the thickness of the composite films could be controlled by the thickness and molecular mass of previously coated polyelectrolyte, treatment time and concentration of particles suspension.

To better understand the structure of nanocomposite coatings by self-assembly method, Figure 7 illustrates the typical schematic morphology of the coatings prepared according to different self-assembly principles.

The three methods described above have been successfully applied in the preparation of O/I nanocomposite coatings, but some problems still exist.

The Problems and Some Advice

For the embedding method, except for nanosilica, it is still not easy to obtain transparent nanocomposite coatings using nanopowder as the nanoparticles. In addition, large volume of nanoparticles may be required to attain the desirable coatings properties for some cases, which increases the cost of coatings since most of nanoparticles are more expensive than resins.

From the viewpoint of the dispersion of nanoparticles, surface modification of nanoparticles or pre-preparation of nanoparticles dispersion may be a good choice. Although some nanoparticle dispersions as additives for coatings are commercially available now, it is regretful that we have not found any nanoparticle slurries with wide applicability yet because of the inherent complexity of traditional coatings caused by its solvents, resins, crosslinking agent and additives. Thus, more kinds of nanoparticle dispersions still need to be exploited. Besides the preparation of nanoparticle dispersion, endowment of nanoparticles with similar surface chemistry to their embedded resin is also important. It can ensure the nanocomposite resin has a comparable miscibility and/or chemical reactivity with the original resin. These kinds of nanoparticles can be prepared by surface modification of nanoparticles with functionalized siloxane and/or further *in-situ* polymerization in corresponding resin precursor. Additionally, nanoparticles, particularly colloidal nanoparticles, may be prepared to organic-inorganic nanocomposite microspheres with controllable morphol- ogies such as raspberry-like, core-shell structure, which can further form high performance coatings (*101-103*).

Sol-gel process is an ideal method for preparation of transparent coatings. Moreover, the properties of sol-gel derived nanocomposite coatings can be easily tuned by various reaction conditions. Currently, combining sol-gel derived hybrid coatings with inorganic nanoparticles to produce high performance coating or provide coating with new functionality is becoming one of the most important research interests for sol-gel derived hybrid coatings.

Self-assembly method is especially suitable for preparation of ordered or special nanostructure coatings and already finds its value in preparation of bionic coatings. At present, self-assembly method is not applicable for the large yield coatings. However, it can be used for surface pre-treatment, micro-device coatings. Generally, self-assembly process is rigorous (for EISA process) or time-consuming (for layer-by-layer process), which restrict its practical application. The studies on self-assembly derived coatings are not enough and should be further conducted in the future.

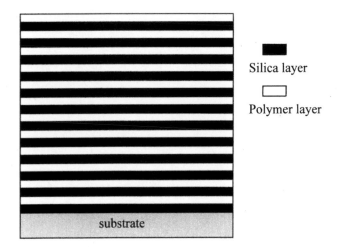

(a) The lamellar structure of coating from EISA process

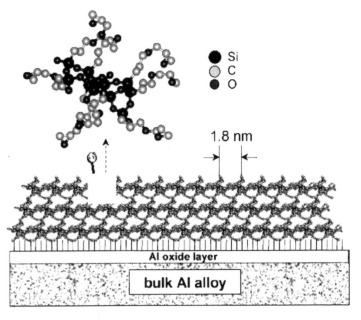

(b) The self-organization of silica cluster of coating from SNAP process
(Reproduced with permission from reference 88. Copyright 2003 Elsevier B.V.)

Figure 7. Schematic diagram of the typical structure of nanocomposite coatings prepared according to different self-assembly principle

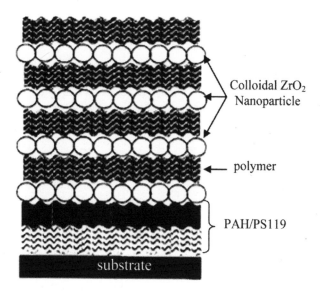

Colloidal ZrO$_2$
Nanoparticle

polymer

PAH/PS119

substrate

Figure 7. Continued. (c) The layer structure of coating from ESSA process (Reproduced with permission from reference 97. Copyright 1998 WILEY-VCH Verlag GmbH.)

Conclusions

O/I nanocomposite coatings are gradually becoming one of the most stirring areas in coatings. Three main methods have been developed to prepare nanocomposite coatings so far. However, the studies on the relationship between preparation, structure and properties of nanocomposite coatings are not sufficient. A universal prediction of the structure and properties of nanocomposite coatings is nearly possible because of the complexity of coating formulations. Currently, there is no doubt that nanocomposite coatings have been finding their usefulness as transparent abrasion and scratch resistant coatings, high weather resistant coatings even functional coatings. It is believed that more novel methods or more new uses will be developed for this new type of coating in the near future.

Acknowledgement

This work is financially supported by Shanghai Natural Science Foundation (Contract No. 04ZR14017) and Shanghai Nano Special Foundation.

References

1. http://www.benzworld.org/news/news.asp?id=254, "World premiere at Mercedes-Benz: Innovative nano-particle clearcoat offers significantly greater scratch resistance and improved gloss", December 3, 2003.
2. http://www.nanopowderenterprises.com/hard.html. "Hard ceramic-like coatings".
3. Fernando, R. *JCT Coatingstech.* **2004**, *5*, 32-38.
4. Baer, D. R.; Burrows, P.E.; El-Azab, A. A. *Prog. Org. Coat.* **2003**, *47*, 342-356.
5. Zhou, S. X.; Wu, L. M.; Sun, J.; Shen, W. D. *Prog. Org. Coat.* **2002**, *45*, 33-42.
6. Zhou, S. X.; Wu, L. M.; Sun, J.; Shen, W. D. *J. Appl. Polym. Sci.* **2003**, *88*, 189-193.
7. Zhang, M. Q.; Rong, M. Z.; Yu, S. L.; Wetzel, B.; Friedrich, K. *Macromol. Mater. Eng.* **2002**, *287*, 111-115.
8. Petrovicova, E.; Knight, R.; Schadler, L. S.; Twardowski, T. E. *J. Appl. Polym. Sci.* **2000**, *78*, 2272-2289.
9. Glasel, H. J.; Bauer, F.; Ernst, H.; Findeisen, M.; Hartmann, E.; Langguth, H.; Mehnert, R.; Schubert, R. *Macromol. Chem. Phys.*, **2000**, *201*, 2765-2770.
10. Zhou, S. X.; Wu, L. M.; Shen, W. D.; Gu, G. X. *J. Mater. Sci.* **2004**, *39*, 1593-1600.
11. Chen, X. C.; Wu, L. M.; Zhou, S. X.; You, B. *Polym. Int.* **2003**, *52*, 993-998.
12. Soloukhin, V.A.; Posthumus, W.; Brokken-Zijp, J. C. M.; Loos, J.; With, G. *Polymer*, **2002**, *43*, 6169-6181.
13. Chen, G. D.; Zhou, S.X.; Wu, L. M. *J. Colloid Interf. Sci.* **2005**, *281*, 339-350.
14. Chen, Y. C.; Zhou, S. X.; Yang, H. H.; Wu, L. M. *J. Appl. Polym. Sci.*, **2005**, *95*, 1032-1039.
15. Bauer, F.; Glasel, H. J.; Decker, U.; Ernst, H.; Freyer, A.; Hartmann, E.; Sauerland, V.; Mehnert, R. *Prog. Org. Coat.* **2003**, *47*, 147-153.
16. Glasel, H. J.; Hartmann, E.; Mehnert, R.; Hirsch, D.; Bottcher, R.; Hormes, J. *Nucl. Instr. and Meth. B*, **1999**, *151*, 200-206.
17. Mennig, M.; Fries, K.; Lindenstruth, M.; Schmidt, H. *Thin Solid Films* **1999**, *351*, 230-234.
18. Bauer, F.; Sauerland, V.; Glasel, H. J.; Ernst, H.; Findeisen, M.; Hartmann, E.; Langguth, H.; Marquardt, B.; Mehnert, R. *Macromol. Mater. Eng.* **2002**, *287*, 546-552.
19. http://www.pcimag.com/, "Nanoparticle Composites for Coating Applications", Paint & Coatings Industry, May, 2004.
20. Jia, Z. Q.; Du, Z. X.; Chen, J. F.; Rao, G. Y.; Wang, Y. H. *Mater. Sci. Eng.*, **2000**, *18*, 100-102 (in Chinese).
21. Yan, L. L.; Wang, K.; Ye, L. *Mater. Sci. Lett.* **2003**, *22*, 1713-1717.
22. Xiong, M. N.; Gu, G. X.; You, B.; Wu, L. M. *J. Appl. Polym. Sci.*, **2003**, *90*, 1923-1931.
23. Li, H. Y.; Chen, Y. F.; Ruan, C. X.; Gao, W. M.; Xie, Y. S. *J. Nanoparticle Res.* **2001**, *3*, 157-160.
24. Xu, T.; Xie, C. S. *Prog. Org. Coat.*, **2003**, *46*, 297-301.

25. Allen, N. S.; Edge, M.; Ortega, A.; Liauw, C. M.; Stratton, J.; McIntyre, R. B. *Polym. Degrad. Stabil.* **2002**, *78*, 467-478.
26. Allen, N. S.; Edge, M.; Ortega, A.; Sandoval, G.; Liauw, C. M.; Verran, J.; Stratton, J.; McIntyre, R. B. *Polym. Degrad. Stabil.* **2004**, *85*, 927-946.
27. He, Q. Y.; Wu, L. M.; Gu, G. X.; Bo, Y. *High Performance Polymer* **2002**, *14*, 383-396
28. Zhou, S. X.; Wu, L. M.; Xiong, M. N.; He, Q. Y.; Chen, G. D. *J. Disper. Sci. Tech.* **2004**, *25*, 417-433.
29. Sugama, T.; Gawlik, K. *Polym. & Polym. Compos.* **2004**, *12*, 153-167.
30. Sun, J.; Gerberich, W. W.; Francis, L. F. *J. Polym. Sci. Part B: Polym. Phys.* **2003**, *41*, 1744-1761.
31. Liu, Y. L.; Yu, Z. F.; Zhou, S. X.; Wu, L. M. *J. Disper. Sci. Tech.* **2006**, *27*, 983-990
32. Manfred, B.; Theodor, E.; Stefan, G.; Bernd, K.; Philip, Y.; Gerhard, J.; Ulrike, D. U.S. Patent 6,020,419, 1998.
33. Schadler, L. S.; Laul, K. O.; Smith, R. W.; Petrovicova, E. *J. Therm. Spray Technol.* **1997**, *6*, 475-485.
34. Petrovicova, E.; Knight, R.; Schadler, L.S.; Twardowski, T.E. *J. Appl. Polym. Sci.* **2000**, *77*, 1684-1699.
35. Chen, X. C.; You, B.; Zhou, S. X.; Wu, L..M. *Surf. Interface Anal.* **2003**, *35*, 369-437.
36. Chen, G. D.; Zhou, S. X.; Liao, H. M.; Wu, L. M. *J. Compos. Mater.* **2005**, *39*, 215-231.
37. Bauer, F.; Ernst, H.; Decker, U.; Findeisen, M.; Glasel, H. J.; Langguth, H.; Hartmann, E.; Mehnert, R.; Peuker, C. *Macromol. Chem. Phys.* **2000**, *201*, 2654-2659.
38. Douce, J.; Boilot, J. P.; Biteau, J.; Scodellaro, L.; Jimenez, A. *Thin Solid Films* **2004**, *466*, 114-122
39. Bauer, F.; Ernst, H.; Hirsch, D.; Naumov, S.; Pelzing, M.; Sauerland, V.; Mehnert, R. *Macromol. Chem. Phys.* **2004**, *205*, 1587-1593.
40. Yeh, J. M.; Liou, S. J.; Lai, C. Y.; Wu, P. C. *Chem. Mater.* **2001**, *13*, 1131-1136.
41. Yeh, J. M.; Chen, C. L.; Chen, Y. C.; Ma, C. Y.; Lee, K. R.; Wei, Y.; Li, S. X. *Polymer* **2002**, *43*, 2729-2736.
42. Yeh, J. M.; Liou, S. J.; Lin, C. Y.; Cheng, C. Y.; Chang, Y. W. *Chem. Mater.* **2002**, *14*, 154-161.
43. Yeh, J. M.; Liou, S. J.; Lu, H. J.; Huang, H. Y. *J. Appl. Polym. Sci.* **2004**, *92*, 2269-2277.
44. Yeh, J. M.; Chen, C. L.; Chen, Y. C.; Ma, C. Y.; Huang, H. Y.; Yu, Y. H. *J. Appl. Polym. Sci.* **2004**, *92*, 631-637.
45. Yu, Y. H.; Yeh, J. M.; Liou, S. J.; Chang, Y. P. *Acta Mater.* **2004**, *52*, 475-486.
46. Shemper, B. S.; Morizur, J. F.; Alirol, M.; Domenech, A.; Hulin, V.; Mathias, L. *J. Appl. Polym. Sci.* **2004**, *93*, 1252-1263.
47. Chen, C. G.; Khobaib, M.; Curliss, D. *Prog. Org. Coat.* **2003**, *47*, 376-383.
48. Majumdar, D.; Blanton, T. N.; Schwark, D. W. *Appl. Clay Sci.* **2003**, *23*, 265-273.
49. Ranade, A.; DSouza, N. A.; Gnade, B. *Polymer* **2002**, *43*, 3759-3766.

50. Mackenzie, J. D.; Bescher, E.P. *J. Sol-Gel Sci. Technol.* **2000**, *19*, 23-29.
51. Schottner, G.; Rose, K.; Posset, U. *J. Sol-Gel Sci. Technol.* **2003**, *27*, 71-79.
52. Sam, E.D.; Arpat, E.; Gunay, V. *Euro. Ceramics VIII, Pts1-3 Key Engineering Materials* **2004**, *264-268*, 395-398.
53. Schmidt, H. *J. Non-cryst. Solids* **1994**, *178*, 302-312.
54. Que, W. X.; Zhang, Q. Y.; Chan, Y. C.; Kam, C. H. *Compos. Sci. Technol.* **2003**, *63*, 347-351.
55. Messori, M.; Toselli, M.; Pilati, F.; Fabbri, P.; Busoli, S.; Pasquali, L.; Nannarone, S. *Polymer* **2003**, *44*, 4463-4470.
56. Chou, T. P.; Chandrasekaran, C.; Limmer, S. J.; Seraji, S.; Wu, Y.; Forbess, M. J.; Nguyen, C.; Cao, G. Z. *J. Non-Cryst. Solids* **2001**, *290*, 153-162.
57. Conde, A.; Duran, A.; Damborenea, M. *Prog. Org. Coat.* **2003**, *46*, 288-296.
58. Li, G. Z.; Wang, L. C.; Ni, H. L.; Pittman, C. U. *J. Inorg. Organomet. Polym.* **2001**, *11*, 123-253.
59. Chen, H.; Zhou, S.X.; You, B.; Wu, L.M. *Mater. Sci. Eng.* **2006**, *24(1)*, 49-52 (in Chinese).
60. Tsujimoto, T.; Uyama, H.; Kobayashi, S. *Macromol. Rapid Comm.* **2003**, *24*, 711-714.
61. Sepeur, S.; Kunze, N.; Werner, B.; Schmidt, H. *Thin Solid Films* **1999**, *351*, 216-219.
62. Spirkova, M.; Brus, J.; Hlavata, D.; Kamisova, H.; Matejka, L.; Strachota, A. *J. Appl. Polym. Sci.* **2004**, *92*, 937-950.
63. Hwang, D. K.; Moon, J. H.; Shul, Y. G.; Jung, K. T.; Kim, D. H.; Lee, D. W. *J. Sol-gel Sci. Technol.* **2003**, *26*, 783-787.
64. Chen, Y. F.; Jin, L. M.; Xie, Y. S. *J. Sol-gel Sci. Technol.* **1998**, *13*, 735-738.
65. Robertson, M. A.; Rudkin, R. A.; Parsonage, D.; Atkinson, A. *J. Sol-gel Sci. Technol.* **2003**, *26*, 291-295.
66. Harreld, J. H.; Esaki, A.; Stucky, G. D. *Chem Mater.* **2003**, *15*, 3481-3489.
67. Dworak, D. P.; Soucek, M. D. *Prog. Org. Coat.* **2003**, *47*, 448-457.
68. Gilberts, J.; Tinnemans, A. H. A.; Hogerheide, M.P.; Koster, T. P. M. *J. Sol-gel Sci. Technol.* **1998**, *11*, 153-159.
69. Xiong, M. N.; You, B.; Zhou, S. X.; Wu, L. M. *Polymer* **2004**, *45*, 2967-2976.
70. Mao, L. Q.; Ding, Y.; Zhang, Z. J.; Dang, H. X. *J. Inorg. Mater.* **2004**, *19*, 391-396(in Chinese).
71. Zaharescu, M.; Crisan, M.; Szatvanyi, A.; Gartner, M. *J. Optoelectr. Adv. Mater.* **2000**, *2*, 618-622.
72. Nguyen, V.; Perrin, F. X.; Vernet, J. L. *Mater. and Corros.-Werkstoffe Und Korrosion* **2004**, *55*, 659-664.
73. Perrin, F. X.; Nguyen, V.; Vernet, J. L. *Polymer* **2002**, *43*, 6159-6167.
74. Kim, T. Y.; Alhooshani, K.; Kabir, A.; Fries, D. P.; Malik, A. *J. Chromatography* **2004**, *1047*, 165-174.
75. Yeh, J. M.; Weng, C. J.; Huang, K. Y.; Huang, H. Y.; Yu, Y. H.; Yin, C. H. *J. Appl. Polym. Sci.* **2004**, *94*, 400-405.
76. Xiong, M. N.; Zhou, S. X.; Wu, L. M.; Wang, B.; Yang, L. *Polymer*, **2004**, *45*, 8127-8138.
77. Chen, W. C.; Lee, S. J.; Lee, L. H.; Lin, J. L. *J. Mater. Chem.* **1999**, *9*, 2999-3003.

78. Yuwono, A. H.; Xue, J. M.; Wang, J.; Elim, H. I.; Ji, W.; Li, Y. T.; White, J. *J. Mater. Chem.* **2003**, *13*, 1475-1479.
79. Sarwar, M. I.; Ahmad, Z. *Eur. Polym. J.* **2000**, *36*, 89-94.
80. Akamatsu, Y.; Makita, K.; Inaba, H.; Minami, T. *J. Ceram. Soc. Jpn.* **2000**, *108*, 365-369.
81. Roux, S.; Audebert, P.; Pagetti, J.; Roche, M. *J. Sol-Gel Sci. Technol.* **2003**, *26*, 435-439.
82. Spirkova, M.; Stejskal, J.; Prokes, J., *Macromol. Symp.* **2004**, *212*, 343-348.
83. Deffar, D.; Teng, G. H.; Soucek, M. D. *Macromol. Mater. Eng.* **2001**, *286*, 204-215.
84. Sellinger, A.; Weiss, P. M.; Nguyen, A.; Lu, Y. F.; Assink, R. A.; Gong, W. L.; Brinker, C. J. *Nature* **1998**, *394*, 256-260.
85. Lu, Y. F.; Yang, Y.; Sellinger, A.; Lu, M. C; Huang, J. M.; Fan, H. Y.; Haddad, R.; Lopez, G.; Burns, A. R.; Sasaki, D. Y.; Schelnutt, J.; Brinker, C. J. *Nature* **2001**, *410*, 913-917.
86. McCaughey, B.; Costello, C.; Wang, D. H.; Hampsey, J. E.; Yang, Z. Z.; Li, C. J.; Brinker, C. J.; Lu, Y. F. *Adv. Mater.* **2003**, *15*, 1266-1269.
87. Smarsly, B.; Garnweitner, G.; Assink, R.; Brinker, C. J.; *Prog. Org. Coat.* **2003**, *47*, 393-400.
88. Donley, M. S.; Mantz, R. A.; Khramov, A. N.; Balbyshev, V. N.; Kasten, L. S.; Gaspar, D. J. *Prog. Org. Coat.* **2003**, *47*, 401-415.
89. Khramov, A. N.; Balbyshev, V. N.; Voevodin, N. N.; Donley, M. S. *Prog. Org. Coat.* **2003**, *47*, 207-213.
90. Balbyshev, V. N.; Anderson, K. L.; Sinsawat, A.; Farmer, B. L.; Donley, M. S. *Prog. Org. Coat.* **2003**, *47*, 337-341.
91. Kasten, L. S.; Balbyshev, V. N.; Donley, M. S. *Prog. Org. Coat.* **2003**, *47*, 214-224.
92. Vreugdenhil, A. J.; Balbyshev, V. N.; Donley, M. S. *J. Coat. Technol.* **2001**, *73*, 35-43.
93. Voevodin, N. N.; Balbyshev, V. N.; Khobaib, M.; Donley., M. S. *Prog. Org. Coat.* **2003**, *47*, 416-423.
94. Brus, J.; Spirkova, M.; Hlavata, D.; Strachota, A. *Macromolecules* **2004**, *37*, 1346-1357.
95. Caruso, R. A.; Susha, A.; Caruso, F. *Chem. Mater.* **2001**, *13*, 400-409.
96. Caruso., F.; Caruso, R. A.; Mohwald, H. *Science* **1998**, *282*, 1111-1114.
97. Rosidian, A.; Liu, Y. J.; Claus, R. O.; *Adv. Mater.* **1998**, *10*, 1087-1091.
98. Pavoor, P. V.; Gearing, B. P.; Gorga, R. E.; Bellare, A.; Cohen, R. E. *J. Appl. Polym. Sci.* **2004**, *92*, 439-448.
99. Zhai, L.; Cebeci, F. C.; Cohen, R. E.; Rubner, M. F. *Nano Lett.* **2004**, *4*, 1349-1353.
100. Hattori, H. *Thin Solid Films* **2001**, *385*, 302-306.
101. Tiarks, F.; Landfester, K.; Antonietti, M. *Langmuir* **2001**, *17*, 5775-5780.
102. Chen, M.; Wu, L.M.; Zhou, S.X.; You, B. *Macromolecules* **2004**, *37*, 9613-9619.
103. Zhang, S. W.; Zhou, S. X.; Weng, Y. M.; Wu, L. M. *Langmiur*, **2005**, *21*, 2124-212

Chapter 11

Effects of Alumina and Silica Nanoparticles on Automotive Clear-Coat Properties

Bryce R. Floryancic, Lucas J. Brickweg, and Raymond H. Fernando

Polymers and Coatings Program, Department of Chemistry and Biochemistry, California Polytechnic State University, San Luis Obispo, CA 93407 (www.polymerscoatings.calpoly.edu)

Effects of commercially available alumina and silica nano-particles on automotive refinish polyurethane clear coatings are discussed in this report. Significant improvements in scratch resistance were observed in coatings containing low levels of alumina nanoparticles. Silica particles caused only slight improvements under the test conditions of this study. AFM analysis of coatings indicates that the particles are well dispersed. Effects of silica and alumina nanoparticles on film clarity and adhesion of the coatings to metal substrates are also discussed.

Introduction

Nanomaterial technology, a subset of more broadly defined nanotechnology, is of great interest to coatings industry today. This interest is clearly evident if one examines the number of nanomaterial related research publications and patents, and nanomaterial companies appeared during recent years (1). The primary reason for the high level of interest is the promise of this technology to deliver breakthrough coating performance in a number of areas (e.g., scratch and mar resistance, barrier properties such as corrosion resistance, and mechanical properties).

Benefits of reducing the particle size of latexes used in coatings were recognized by coating scientists several decades ago. Smaller particle size latexes have been shown to provide superior pigment binding ability in architectural paints (2, 3). This understanding led to the introduction of small particle (100 nm and smaller) latexes by raw material suppliers. PUDs (polyurethane dispersions) that are commonly used in coatings provide another example of conventional organic nanoparticles (4). Early use of colloidal silica and other nanoparticles in coatings has been reported as well (5). However, such early work on applications of nanomaterials in coatings has been sporadic and disjointed. Today, there are numerous groups around the world investigating many different aspects of nanomaterial applications in coatings. As a result, conditions are mature for potential breakthroughs in this area. Potential benefits of alumina, silica, and other inorganic nanoparticles in coatings were the focus of several presentations made at a recent conference (6). In this paper, results of a study to determine the effects of alumina and silica nanoparticles on the performance of a polyurethane automotive refinish coating are reported.

Potential Benefits of Inorganic Nanoparticles in Coatings

The basic purpose of applying nano-material technology to coatings is to produce either a nano-composite (e.g., inorganic nano-particles in an organic matrix), or nano-structure within a single phase coating. The category of nanomaterials includes any material having at least one of its dimensions in the range between few nanometers and about 100 nm. The smaller size offers two key features -- high interfacial material content and optical clarity -- both of which are important in automotive coating applications.

1. Interfacial Material Content

It is a well established fact that the properties of "interfacial material" at the interface between two materials are different from the bulk properties of each material. An example is the modification of glass transition temperature (Tg) of a polymer at an interface. Tg of the polymer would change due to the steric and enthalpic effects that alter the segmental mobility of the polymer molecule at a polymer/inorganic filler interface (7-9). Thickness of the interfacial layer can also vary, but typically is in the range of 1 - 10 nm.

Consider an example where an inorganic material made up of uniform, spherical particles is incorporated into an organic coating (Figure 1). The particles are large in one case, and small in the other. In each case there is an interfacial region where several molecular layers of the two phases behave differently from the bulk. Dispersing nano-particles, instead of larger particles allows a coating formulator to increase the interfacial material content significantly. This is illustrated in Table I in which the effect of particle size and the interfacial layer thickness on the volume fraction of interfacial material content is presented. The calculations apply for a dispersion of particles in a continuous polymer matrix at 30 volume% loading. For example, at a particle size of 300 nm and interfacial layer thickness of 10 nm, the interfacial material content is 3%, whereas it increases to 22% when the particle size is decreased to 50 nm. Thus, the interfacial material becomes a major portion of the composite coating. If the interfacial material has better properties that are not offered by individual materials of the composite, this approach would maximize such benefits.

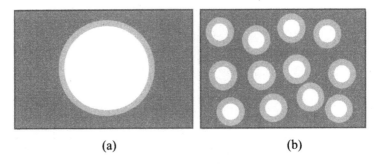

(a)   (b)

Figure 1. Incorporation of Uniform, Spherical Filler Particles in to an Organic Coating: (a) Macro-particles; (b) Nano-particles

What was discussed so far is enhancing interfacial effects that are already present in a macro-composite by transforming the system into nanocomposites. A related but different effect is that even in the absence of entropic or enthalpic effects at the interface, a nanomaterial would have altered properties because a large fraction of the material could be under unbalanced atomic scale forces. Bulk properties of a material are not scalable down to this size scale (referred as non-scalable region (10)).

Table I. Interfacial Material Volume Fraction Dependence on Particle Size
[Particle Loading – 30% by Volume; Interfacial Layer Thickness – 10 nm]

Particle Diameter (nm)	Interfacial Volume Fraction
300	0.03
250	0.04
200	0.05
150	0.06
100	0.10
50	0.22

2. Optical Clarity

Optical clarity is an essential property of a number of classes of coatings such as automobile clear coats, floor wear layers, and optical lens coatings. Unless the refractive index can be matched to that of the coating resin, adding inorganic particles cause light scattering that leads to reduction or elimination of film clarity. Since nanoparticles are much smaller in size compared to the wavelength range of visible light (400 – 800nm), they scatter very little light. Therefore, they can be added to a clear coating formulation with little or no adverse impact on visual characteristics. This feature of nanoparticles is extremely important in expanding nanoparticle applications in coatings.

A critical requirement for reaping the potential benefits of incorporating nanoparticles in a coating is nanoscale dispersion of the particles. In fact, the dispersion issue is one of the major barriers to more rapid introduction of new products in this area. One approach to achieve nano-scale dispersion is to use effective grinding methods such as ball milling (11, 12). Effective dispersing requires grinding media that is much smaller than those used for dispersing conventional particles. The high surface area generated can significantly increase the dispersant demand. A recent article described how to optimize dispersion methods for nanopowders in coatings (12). High viscosity caused by finely dispersed nanoparticles is another problem that needs to be addressed.

The large surface area can increase viscosities due to the increase in interfacial forces and limit the amount of nanoparticles that can be incorporated. Adding the right surface functionality to address dispersibility and viscosity rise is another approach to address dispersion issue. In addition to promoting dispersion, functionalizing the particle surface enables the nanoparticle to be covalently linked to the organic resin matrix.

Most of the nanoparticle research and development in raw material suppliers' laboratories have been focused on functionalizing and other surface treatments of nanoparticles. Such efforts are facilitating commercialization of increasing number of nanoparticle types for coating applications (6). More and more types of nanoparticles are now becoming available with functional groups attached to the particle surface to improve dispersion and enable the particles to be covalently linked with the organic matrix. Several studies have reported (1,6,13-16) the use of functionalized nanoalumina and nanosilica in UV curable resins to prepare UV cured coatings with improved abrasion, scratch, and chemical resistance while maintaining a high gloss and film clarity. In these efforts, high nanoparticle loadings have been achieved by the selection of surface functional groups compatible with the dispersion medium (solvent and/or monomer) to minimize viscosity.

A European luxury car manufacturer recently announced use of nanoparticles (referred to as ceramic particles) in a clear coat application (17). The particles are reported to be less than 20 nm in diameter, and cross-linked with the coating at 140 °C. Compared to conventional clear coats, the new coating system is claimed to have 40% better gloss retention after car wash tests. A few months earlier, a major US paint company announced introduction of a nanoparticle based automotive clear coat with the trade name "CeramiClear". At least two recent patent applications relate to this type of approach (18,19).

Fumed silica represents another important nanoparticle silica technology. This material, made by flame hydrolysis of silicon tetrachloride, has also been around for many decades. However, this process produce fused or agglomerated nano silica particles, and as a result, dispersion is difficult and cause high viscosities at low loading levels. Its primary use in coatings has been as a rheology modifier. However, recently reported results on functionalized fumed silica claim measurable improvements in abrasion, scratch and mar improvements in UV cured clear coats (20)

Experimental

Alumina and silica nanoparticle samples were provided by BYK-Chemie Company. Properties of the nanoparticle samples are given in Table II. An ambient-cure, two-component, polyurethane automotive refinish coating was used as the base formulation in all experiments.

The nanoparticle dispersions were added to the automotive refinish formulation under agitation, and agitation was continued for ten minutes after the addition. Agitation was accomplished with an EIGER Mixer rotating at 1500 rpm. The nanoparticle dispersions dispersed rapidly into the resin and appeared to stay well dispersed. Solvent levels were adjusted to maintain the solids level the same, and therefore the film thickness the same for each coating film.

Table II. Nanoparticles Used in the Study

Type	Medium	% Solids	Particle Size (nm)
Aluminum A	DPnB	38.5	30-40
Aluminum B	TPGDA	40.9	30-40
Aluminum C	Methoxypropyl acetate	32.0	30-40
Silica A	Methoxypropyl acetate/methoxy propanol	32.0	

Each coating sample, except those prepared for gloss testing, was applied on cold rolled steel panels. A 3 mil (76 micron) drawdown bar was used to apply the films. Although the formulation can be cured under ambient conditions, the panels were baked at 70°C for 30 minutes to provide a more standard curing environment and to minimize contamination of the films by dust. Pendulum hardness of cured coatings, stored under ambient conditions, was monitored over a period of 7 days for each nanoparticle system. Scratch resistance was tested by performing 50 double rubs using 0000 grade steel wool. Samples were randomized in order to eliminate operator bias. Weights of the coated panels were measured before and after the scratching process to determine the weight loss due to scratching. Gloss was measured over coatings prepared on black and white primed steel panels. Flexibility of the coatings was tested using the Conical Mandrel Tester (ASTM D 522-93a), and the adhesion of each coating sample was tested using the cross hatch adhesion test (ASTM D 3359-02).

Differential Scanning Calorimetry was performed after aging the coatings for 7 days using a TA Instruments Q1000 DSC. Samples were allowed to equilibrate at 0°C, and then heated to 70°C at a heating rate of 10°C/min. The glass transition temperature was then calculated using the Tg tool in TA Universal Analysis software. Atomic Force Microscopy testing was conducted using a Pacific Nanotechnology AFM. Samples were drawn down at 3 mil wet thickness on glass microscope slides and cured at 70°C for 30 minutes. Testing was conducted in close contact mode with a silicon cantilever using the "C" tip.

Results

Scratch Resistance

Initial scratch testing results are shown in Figure 2. All nanoparticle systems show an initial and overall reduction in coating weight when compared to the control having no added nanoparticles. Silica A showed less reduction in weight lost. The nanoparticles were then formulated into coatings at levels of 0 to 3.33 dry weight percent in intervals of 0.33 % to study the effects in greater detail. The new coating samples were tested by the same scratch testing method. Alumina A caused a significant decrease in weight lost at 0.33 wt.% loading, and then the decrease continued as the nanoparticle loading increased (Figure 3). Behavior of Alumina B containing coatings (Figure 4) was similar to those containing Alumina A. Alumina C showed much more dramatic results. The control coating lost 18 mg, while the coating with 0.33% alumina lost only 5 mg (Figure 5). This initial drop was the largest of all 4 systems. Silica A showed a drop that was less dramatic than all the alumina systems (Figure 6). Silica appears to be less effective at reducing the weight loss than its alumina counterparts under the test conditions of this study.

Haze

All four nanoparticle systems were drawn down onto glass plates and cured, and their haze was observed. A 0-10 rating scale was developed to quantify the haze, 0 being transparent and 10 being opaque. Although the films were all fairly clear, there were some differences in the haze values. Alumina A at a loading of 2.5% had a haze value of 1, while Alumina B and alumina C (at a loading of 3%) had ratings of 2 and 0, respectively (Table III). Alumina C caused no visually detectable haze. Silica A caused no haze even at loadings up to 8%. Silica was expected to not cause haze, as its refractive index is closer to that of the polyurethane. The variance in haze among the alumina systems was not expected, and may be due to differences in particle size distribution. The particles are of the same average size, but Alumina C has a more narrow size distribution (21).

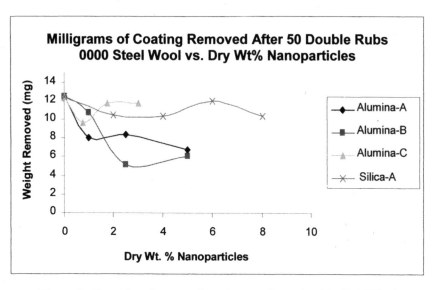

Figure 2. Scratch resistance of coatings as determined by Steel Wool Double Rub test

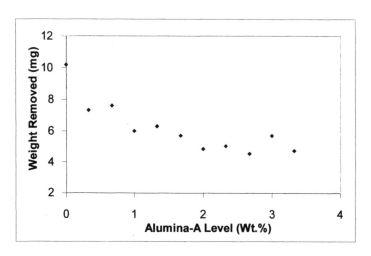

Figure 3. Scratch resistance of coatings containing Alumina A, as determined by Steel Wool Double Rub test

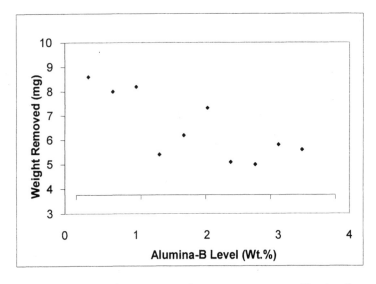

Figure 4. Scratch resistance of coatings containing Alumina B, as determined by Steel Wool Double Rub test

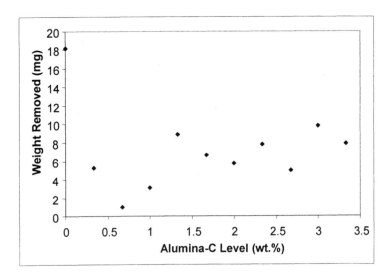

Figure 5. Scratch resistance of coatings containing Alumina C, as determined by Steel Wool Double Rub test

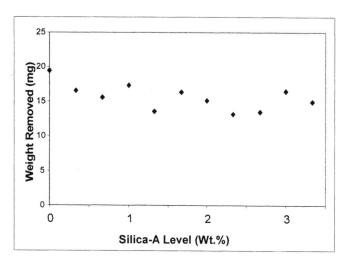

*Figure 6. Scratch resistance of coatings containing Silica A,
as determined by Steel Wool Double Rub test*

Table III. Haze rating for each coating
[0 – transparent; 10 – opaque]

Sample	Haze Rating
No Nanoparticles	0
1% Alumina-A	0
2.5% Alumina-A	1
5% Alumina-A	2
1% Alumina-B	1
2.5% Alumina-B	2
5% Alumina-B	3
0.75% Alumina-C	0
1.75% Alumina-C	0
3% Alumina-C	0
2% Silica-A	0
4% Silica-A	0
6% Silica-A	0
8% Silica-A	0

Adhesion and Mandrel Bend Tests

Cross-hatch adhesion test results are shown in Table IV. The control coating films completely failed the cross hatch adhesion test. Alumina A caused a slight improvement in adhesion, and Alumina B had no effect. Alumina C increased the adhesion of the coating from 0B to a perfect 5B at levels as low as 0.75%. Silica A showed an increase in adhesion with increasing nanoparticle load from 0B at 2% to 5B at 8%. The addition of nanoparticles to the coating was not expected to affect adhesion, and the increase in adhesion is believed to be due to the differences in surface treatment of the particles. All coatings passed the conical Mandrel bend test without cracking or delaminating.

Table IV. Cross-Hatch Adhesion Test Results

Sample	Adhesion Rating
No Nanoparticles	0B
1% Alumina-A	0B
2.5% Alumina-A	0B
5% Alumina-A	1B
1% Alumina-B	0B
2.5% Alumina-B	0B
5% Alumina-B	0B
0.75% Alumina-C	5B
1.75% Alumina-C	5B
3% Alumina-C	5B
2% Silica-A	0B
4% Silica-A	3B
6% Silica-A	4B
8% Silica-A	5B

Differential Scanning Calorimetry

Differential scanning calorimetry was used to determine how the nanoparticles would affect the glass transition temperature of the films. The glass transition temperatures of the coatings containing Alumina A, Alumina C, and Silica A were fairly constant at about 50°C (Figure 7). Alumina-B, however, had a major effect on the Tg of the coating, decreasing to as low as 17°C at higher loadings.

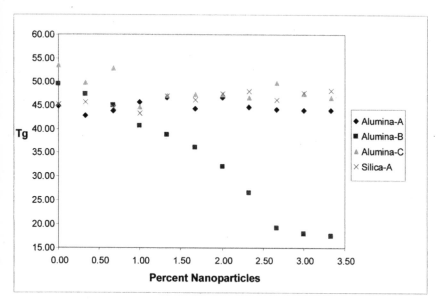

Figure 7. Dependence of glass transition temperature of coatings on nanoparticle content

Pendulum Hardness

Pendulum Hardness tests were performed in the Koenig mode; average dampening times of three trials were recorded. Koenig hardness values of coatings (expressed in seconds) measured after 7 days of aging are shown in Figure 8. The control coatings show average pendulum hardness of about 200 seconds. Alumina-A, Alumina-B, and Silica-A show a slight drop in hardness over the range of nanoparticle loads to about 175 seconds. Alumina-B shows a significant drop from about 175 seconds at 0.33 wt.% loading to about 25 seconds at 3.33 wt.%. The addition of Alumina-B causes the coating to become quite soft at relatively low levels. Not surprisingly, the pendulum hardness data closely resemble the glass transition temperature data.

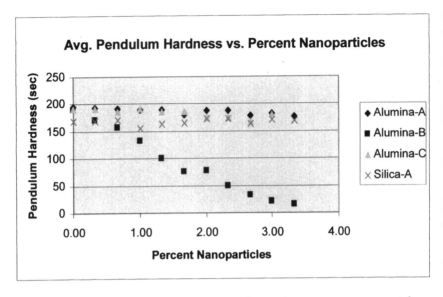

Figure 8. Dependence of Pendulum Hardness of coatings on nanoparticle content

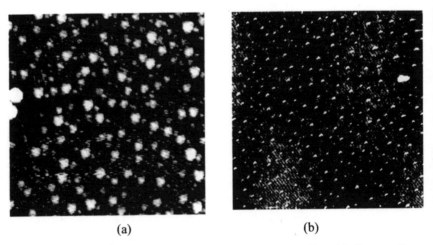

(a)   (b)

Figure 9. AFM Images of coatings containing nanoparticles: (a) Alumina-B at 2.67 wt.% loading – scan area 5x5 micron, (b) Silica-A at 0.67 wt.% loading – scan area 10x10 micron

Atomic Force Microscopy

AFM was used to determine the surface morphology of the films containing nanoparticles. The films were expected to show evenly dispersed nanoparticles on the surface. This is indeed the case for coating AFM images analyzed so far (Figure 9). Detailed AFM analysis of all coating samples is underway.

Discussion

Results of the steel wool scratch damage test clearly indicate improvements caused by alumina nanoparticles. Only slight improvements were observed with silica. This can be explained on the basis of hardness difference of alumina and silica. Alumina A and Alumina C had no significant effect on the T_g of coatings, whereas Alumina C decreased it significantly. However, all three systems showed scratch resistance improvements. The reduction in T_g is most likely due to a plasticizing effect of the tripropyleneglycol diacrylate in which the Alumina B is supplied. In order to test this, a series of coatings with different levels of tripropyleneglycol diacrylate is being prepared. An important aspect of the surface treatment appears to be its effect on adhesion of the coating. Advancements in surface treatment technology are evident in view of the fact that no special dispersing techniques were employed to incorporate the nanoparticles into coatings. AFM data show that the particles are well dispersed.

Best film clarity is achieved with silica nanoparticles. However, there is a significant difference in film clarity among different alumina grades. Although the average particle sizes are similar, Alumina C is reported to have a narrower particle size distribution (21). It appears that elimination of small amounts of larger particles have helped improving the clarity of films made with Alumina C.

Future Work

Future work will focus on efforts to understand detailed mechanisms of scratch resistance improvements. These efforts will be conducted in collaboration with National Institute of Standards and Technology (NIST) and nanomaterial suppliers. These efforts will include controlled scratching of coatings by a Nano-Indenter and determination of particle size distributions of the nanoparticle samples.

Acknowledgements

Authors acknowledge the support of Mr. Robert McMullin and Dr. Thomas Sawitowski of BYK-Chemie for this study. Dr. Derek Gragson and Mr. Ryan Garcia are acknowledged for providing support in Atomic Force Microscopy analysis. Dr. Lipiin Sung (NIST) and Mr. Jeff Comer are collaborating on many aspects of the future work on this project. There collaboration is highly appreciated.

References

1. Fernando, R. H., JCT CoatingsTech, 1(5), 32-38 (2004)
2. Schaller, E. J., Journal of Paint Technology, 40, 433 (1968)
3. Boswell, S. T., Craver, J. K., and Tess, R. W. (Ed.) "Applied Polymer Science", ACS (1975)
4. Klempner, D., and Frisch, K., "Advances in Urethane Science and Technology", University of Detroit, Mercy (2001)
5. Fernando, R. H., and Bohrn, W. J., US 5,124,202 (1992)
6. FSCT Symposium on Nanotechnology in Coatings: Realizing the Potential, Seattle, WA (2005)
7. Tsui, O. K. C., and Russell, T. P, Macromolecules, 34, 5535-5539 (2001)
8. Mayes, A. M., Macromolecules, 27, 3114-3115 (1994)
9. Qiang, X., Zhao, C., JianZun, Y., and Yuan, C. H., Journal of Applied Polymer Science, 91, 2739-2749 (2004)
10. Baer, D. R., Burrows, P. E., and El-Azab, A. A., Progress in Organic Coatings, 47, 342-356 (2003)
11. Koch, C. C., Rev. Adv. Mater. Sci., 5, 91-99 (2003)
12. Way, H. W., JCT CoatingsTech, 1(1), 54-60 (2004)
13. Adebahr, T., Roscher, C., and Adam, J., European Coatings Journal, 4, 144-147 (2001)
14. Lewis, L. N., and Katsamberis, Journal of Applied Polymer Science, 42, 1551 (1991)
15. "Tiny Particles, Huge Effect", Paint & Coatings Industry, Internet Edition, Posted on 10/01/2003
16. "Staining Preventive UV Coating", Paint & Coatings Industry, Internet Edition, Posted on 10/01/2002:
17. "Innovative Nano-Particle Clearcoat Offers Significantly Greater Scratch Resistance and Improved Gloss", http://www.benzworld.org/news/news.asp?id=254, December 3, 2003
18. Vanier, N. R., Munro, C. H., Clarr, J. A., and Jennings, R. E., US 30162015:A1 (2003)

19. Vanier, N. R., Munro, C. H., McCollum, G. J., O'Dwyer, J. B., and Kutchko, C., US 301662876:A1 (2003)
20. "New Modified Silicon Dioxides for Radiation Cured Coatings", DeGussa AG, Technical Bulletin
21. Cayton, R. H., Brotzman, R., Murray, P., Sawitowski, T., Berkei, M., and Nolte, U., Proceedings of ACS Polym. Mat. Sci. Eng. Symposium on Nanotechnology Applications in Coatings, 95, 249 (2006)

Novel Coatings

Chapter 12

Biocatalytic Paints and Coatings

C. Steven McDaniel[1], Jesse McDaniel[1,2], James R. Wild[2], and Melinda E. Wales[1,2]

[1]Reactive Surfaces, Limited, 300 West Avenue, Austin, TX 78701
[2]Department of Biochemistry and Biophysics, Texas A&M University, College Station, TX 77843

Enzymes have been used in a variety of forms, including free in solution and immobilized or crosslinked to solid supports. The conventional notion that enzymes are only active in aqueous media has long been discarded due to the numerous studies documenting enzyme activities in nonaqueous media. The incorporation of enzymes and other proteins into polymeric coatings and films has been investigated in this study with the goal of generating biologically active coating materials. Of all the enzymes, hydrolases are the most employed for industrial biotransformations. It is estimated that approximately 80% of all industrially used enzymes are hydrolases. By tailoring organophosphorus hydrolase (OPH) for specific purposes, coating additives have been bioengineered to exhibit activity and/or specificity for decontaminating chemical warfare agents. The incorporation of enzyme-based additives into paints and coatings should allow for self-cleaning surfaces, mold-inhibiting surfaces, deodorizing surfaces, textile coatings, and catalytic coatings for waste stream decontamination.

Introduction

Organophosphorus compounds (OP) are used extensively as neurotoxic pesticides, and are highly toxic to many different organisms, including humans. The primary effects of exposure to these compounds results from the inhibition of the enzymes acetylcholinesterase and butyrylcholinesterase, leading to the failure of autonomic and central nervous systems (*1*). Over 40 million kilograms of OP pesticides are used in the United States annually (*2*). The number of people accidentally poisoned by OP pesticides has been estimated to be upwards of 500,000 persons a year (*3*). Depending on the OP compound and its toxicity, a repeated, prolonged and/or low-dose exposure to an OP compound can cause neurological dysfunction and delayed cholinergic toxicity. High-dose exposure can be fatal (*4*) or lead to OP-delayed-onset polyneuropathy. Arguably of greater immediate danger to humans, however, is the use of some of the most toxic OP compounds as chemical warfare agents (CWA). Chemical warfare agents are classified into two major categories: G agents [GD[1] (soman), GB (sarin), GF (cyclosarin) or GA (tabun)], and the methyl phosphonothioates commonly known as V agents [VX and Russian VX (R-VX or VR)]. All of the neat CWAs are colorless liquids with varying volatilities. By addition of a thickener (e.g., a variety of carbon polymers), soman or other more volatile agents may be made to be less volatile and more persistent.

The CWAs are extremely toxic and have a rapid effect, entering the body through inhalation, direct contact to the skin with a gas or with a contaminated surface, or through ingestion of contaminated food or drink. Depending on the concentration and route of exposure, the appearance of the first symptoms may be within seconds or as long as 20 to 30 minutes after exposure. In addition, some surfaces that are exposed to the agents may retain their toxicity for long periods of time; the OP nerve agents may penetrate (and/or attach to) materials such as paints, plastics, and rubbers, allowing agents to remain in those materials and be released over time. "Thickened" nerve agents are even more persistent and difficult to decontaminate from surfaces such as walls, vehicles, or personal items such as a computer keyboards, night-vision devises, and weapons. While there remains some inconsistency in the values, Army manuals generally agree that the chemical agents Sarin, Soman, and Mustard Gas present a vapor hazard for one to ten hours, and VX for 241 to 1776 hours (ten to seventy-four days). In addition to weather and agent formulation, the surface plays a role in persistency, with liquid neat agent on asphalt lasting minutes and on unpainted metal or glass, hours to weeks. (*5*)

The most important method of protection from nerve agents is to prevent exposure. For military personnel and other first responders, masks and full body protective gear are available. However, the impermeable suits and even some air permeable suits are bulky and hot, often inhibiting free movement, thus making tasks harder and longer to complete. Tasks requiring detailed handwork, such as use of communication equipment, can be severely hampered by such bulky protective gear. In addition, from a homeland security perspective, it is difficult to provide the general population with such protective equipment.

Historically, most technologies for chemical warfare agent decontamination have focused on the treatment of surfaces after chemical exposure, whether actual or merely suspected, may have occurred. There are several current methods of decontamination of surfaces. One common method involves post-exposure washing with hot water with or without addition of detergents or organic solvents, such as caustic solutions or foams (e.g., Eco, Sandia, Decon Green). Other potentially-effective methods include the use of intensive heat and carbon dioxide applied for sustained periods, or incorporation of oxidizing materials (e.g., TiO_2 and porphyrins) into coatings that, when exposed to sustained high levels of UV light, degrade chemical agents (6). Chemical agent resistant coatings ("CARCs") have been developed to withstand repeated decontamination efforts with such caustic and organic solvents. Although each of these approaches can be effective under specific conditions, a number of limitations exist. Caustic solutions degrade surfaces, create personnel handling and environmental risks, and require transport and mixing logistics. The use of alkaline solutions, such as a bleaching agent, is both relatively slow in chemically degrading VX and can produce decontamination products as toxic as the OP itself (7).

A much studied but underutilized approach to OP decontamination is the application of OP degrading enzymes. Different OP compound degrading enzymes have been described, primarily for the detoxification of OP pesticides (8, 9). However, to date, there has been limited success in harnessing the potential of these enzymes in systems that can be readily and cost effectively used in field-based military or civilian applications. Thus, despite the current understanding of the various OP compound degrading compositions and techniques, whether based on caustic chemicals or enzymes, there is a clear and present need for compositions and methods that can readily be used in OP compound degradation. In particular, there is a clear need for surface coatings that can be prophylactically applied or applied immediately post-contamination in order to quickly decontaminate the surface.

Materials and Methods

Reagents

The OP pesticides paraoxon and demeton-S were obtained from ChemService (Westchester, PA), and diisopropylfluorophosphate (DFP) was obtained from Sigma (St. Louis, MO). All buffer and assay reagents (MES, N-ethylmorpholine, diethanolamine, DTNB, CHES) were also obtained from Sigma-Aldrich (St. Louis, MO). Media components for cell growth (Bacto-Tryptone, Yeast Extract, Bacto-Agar) were obtained from Fisher Scientific

(Pittsburgh, PA), as was Difco™ Terrific Broth (TB). *Escherichia coli* DH5α (*supE44 ΔlacU169 [Φ80 lacZΔM15] hsdR17 recA1 endA1 gryA96 thi-1 relA1*) was used to express the bacterial OPH from the recombinant plasmids pOP419-WT or pOP419-mut (*10*). The plasmid pOP419-WT contains the wild-type *opd* gene which encodes the native OPH without its amino-terminal leader sequence, while pOP419-mut is a derivative of pOP419 which contains a double mutation at the 254 and 257 residues (11). Paints selected as the coating matrix for entrapping enzymatic activity were Sherwin-Williams Acrylic Latex, and Olympic Latex. Testing coupons were manufactured by Q-panel Lab Products (Cleveland, OH), and were 0.032" x 1 cm x 10 cm aluminum flat panels.

Cell Growth and Enzymatic Additive Preparation

OPDtox™ additive of Reactive Surfaces, Ltd (Austin, Texas, USA) was used in all experiments. For the studies described here, OPDtox™ was prepared as a dried whole-cell powder as follows. *E. coli* DH5α transformed with either pOP419-WT or pOP419-mut was plated on LB (10 g. BactoTryptone, 5 g. yeast extract, 5 g. NaOH per liter) plates containing 50 μg/ml ampicillin (Amp) and incubated overnight at 37 °C. A single colony from each transformation was used to inoculate 5 ml tubes containing LB^{amp} media and incubated at 37 °C overnight, or approximately 16 hrs. Ten mls of each was used to inoculate 1 L of TB supplemented with 0.4% glycerol, 50 μg/ml Amp and 1 mM $CoCl_2$, which was grown at 30°C and 185 rpm for approximately 40 hrs, and then harvested by centrifugation at 4000 x g. The cell pellets were twice washed and then resuspended 1:1 (w/v) in ddH_2O, shell frozen in 2 L freeze-dry flasks in a dry ice-ethanol bath, and then lyophilized overnight at -50 °C and between 10 and 100 microns Hg vacuum. Cells that were not lyophilized immediately were shell frozen and stored at -70°C until lyophilization. The lyophilized product was characterized for enzymatic activity and used as an additive (OPDtox™) for the paint and coating experiments. Control additive was made in an identical fashion, with the exception that the *E. coli* DH5α was not transformed, and so was enzymatically inactive for the relevant enzyme, organophoshorus hydrolase.

Surface and Coating Preparation

Prior to the application of coatings, all metal surfaces were degreased using Simple Green, washed with ddH_2O, and dried at 52°C (approximately 30 min). A primer coat was prepared by diluting the latex paint, according to manufacturer's specifications, to 87.5% (v/v) in ddH_2O. Bioactive paints were prepared by first resuspending additive in 50% glycerol. This was added to paint in a 1:10 (v:v) ratio, pre-mixed with ammonium bicarbonate (the final concentration ranged from 0 to 1 M). The final amount of additive per gallon of paint varied in the individual experiments between 100 g/gal and 400 g/gal.

Coatings were evenly applied using a Gardco paint spreader at a thickness of 50 mils, and allowed to air dry for a minimum of 24 hrs.

For wooden surface assays, dowels 45 mm in length and approximately 2 mm in diameter were prepared. The dowels were painted by dipping in the bioactive paint, prepared as described previously, and hung to allow excess paint to drain. For assays, the painted dowels were cut into either 5 or 15 mm lengths. For agent assays, the inside of 12 x 32 mm vials were coated by dispensing approximately 1 ml of bioactive paint into each vial and rolling to coat the inside walls. The vials were allowed to dry for several days before assaying.

Solution Testing of Coated Surfaces

Prepared dowels, either 5 or 15 mm long, were added to a 1.5 mL tube containing 1 mM paraoxon in 2 ml phosphate buffer, pH 8.0. Ten μL samples were taken at 1 minute intervals to monitor the reaction progress by following the production of the cleavage product p-nitrophenol (pNP, $\varepsilon_{400} = 17,000$ M^{-1} cm^{-1}). The reaction was mixed by inversion before each time point. Alternatively, 15 mm dowels were used in a 3 ml reaction vessel. The reagent volumes were increased proportionately. Two sets of control dowels were used, one of which was coated with paint alone, and the second set was coated with paint prepared using the control additive. Activity with DFP as substrate was measured by monitoring the production of fluoride from DFP with a fluoride ion sensitive electrode (12). The assay solution was 1 mM DFP in phosphate buffer, pH 8.0.

To evaluate the ability of the biocatalytic coating to decontaminate chemical agents, the insides of glass vials were coated as described above. The vial was contaminated with 1 μg/ml agent in 1 ml 10 % (v/v) water-in-methanol. The sample was placed in the auto-sampler tray of the GC, and sampled at irregular intervals up to 500 and 600 minutes. Sampling ceased when the change between two consecutive measurements was minimal.

Dry Surface Testing

The OP pesticides used in these tests were paraoxon (PX, $\rho = 1.278$ g/ml) and demeton-S (D-S, $\rho = 1.132$ g/ml). In the standard surface assay, individual test coupons were contaminated with PX, and D-S by drop-dissemination for the required contamination density. The contaminated test coupons were weathered for prescribed residence times of 0 and 1,440 minutes. Each time point was performed in triplicate. Coupons were extracted with isopropanol (IPA) for 1.5 hrs to quench the reaction and remove remaining agent from the surface. Gas chromatographic (GC) analyses were conducted to determine the amount of OP compound that had been hydrolyzed. Parameters which were varied and compared to this standard protocol included contamination density (50 x 0.2 μl droplets, 5 x 2.0 μl droplets, and 5 x 2.0 μl droplets smeared to a thin film) and drying time (4 to 72 hours).

Gas Chromatography

Extracts resulting from of the surfaces assays were quantified on a GC-FPD, with a Phenomenex ZB-5 (30 m x 0.32 mm ID x 0.25 um df; 5% phenyl-95%-dimethylpolysiloxane) column. A splitless injection port was utilized, with helium carrier gas, an initial temperature of 250 °C, and flow rate of 39.6 ml/min. The initial oven temperature was 80 °C for 0.5 min, ramp rate 35 °C/min to 260 °C hold 0.5 min, ramp rate 40 °C/min to 290 °C hold 1 min, with a total run time 7.89 min. The FPD temperature was 225 °C, hydrogen flow 75 ml/min, air flow 100 ml/min, and helium makeup flow 13.4 ml/min. Linearity of the FPD was determined by injecting serial dilutions of each OP in isopropanol.

Results and Discussion

Characterization of Biocatalytic Coatings in Organic Solvent

The ability of the OPDtox™ to retain catalytic activity once embedded into a coating was demonstrated by using wooden dowels with a biocatalytic coating as the catalytic component in assays with paraoxon (Figure 1). The catalytic coating hydrolyzed 80% of a 1 mM (275 ppm) solution of paraoxon within 20 minutes. This is a decontamination rate of 980 nM paraoxon/min/mm^2 surface area. Similarly, with DFP as the substrate, the dowels demonstrated 70% hydrolysis of the DFP by 1 hour and a decontamination rate of 170 nM DFP/min/mm^2. These results were extended to the OP CWA by monitoring hydrolysis of VX, soman (GD), cyclohexyl methylphosphonofluoridate (GF) and the pesticide surrogate DFP in 90% methanol (Figure 2). In spite of significant inhibition of the enzymatic activity by the methanol solvent system (13), the biocatalytic coatings performed well in the decontamination of the test agents. (The activity was determined as % substrate remaining after 24 hrs.) While the mechanism by which methanol specifically affects the OPH catalyzed reactions is undefined, the likelihood that the target substrates would be part of an organic solvent system makes the design of an organic tolerant system a critical part of any widely applicable, enzymatic OP decontamination procedure. These results suggested that the application of a biocatalytic coating provides a significant step toward that goal.

Characterization of Biocatalytic Coatings in Aqueous Solution

To evaluate the potential for surface catalysis in the absence of water or other solvent system, the surfaces were evaluated with a dry surface assay. As can be seen in Figure 3, the hydrolysis is significantly slower than the same surface in a solution decontamination scenario, with a calculated rate of approximately 8 pM/min/mm^2 with paraoxon and slower still with demeton-S.

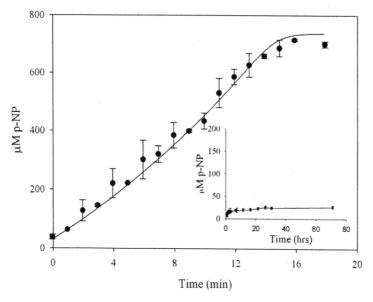

Figure 1. The ability of the biocatalytic coating to decontaminate the pesticide paraoxon is demonstrated. Inset: The control dowel was monitored for more than 60 hrs. to evaluate the potential for hydrolysis of paraoxon by paint alone.

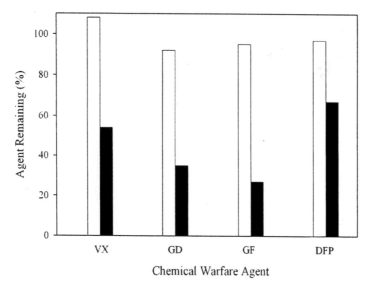

Figure 2. Efficacy of the biocatalytic coatings to hydrolyse CWA. The light bars are the control vials, coated with catalytically inactive paint, the darker bars are the test vials, coated with bioactive paints prepared with OPDtox.

(The coupon series in the figure is as follows: A, negative control, paint + catalytically inactive OPDtox; B, biocatalytic surface with OPDtox targeting paraoxon; C, biocatalytic surface targeting the P-S bond substrate, demeton-S.) Subsequent testing of these surfaces in the solution assay confirmed that the coatings remain catalytically active, which suggested the possibility that the transfer/diffusion of substrate and product on a dry surface is problematic. To further evaluate this possibility, the reactivity of the catalytic surfaces was evaluated as the coating dried. Within 72 hrs, the reactivity of the surface decreases from approximately 50% hydrolysis to less than 10%. This indicated that although the applications of the catalytic additive OPDtox remain active in off-the-shelf paints, their ability to reduce contamination was impacted by their state of hydration. This possibility was confirmed by simply rehydrating the surfaces with water, which resulted in the recovery of greater than 85% of the original activity (Figure 4). To distinguish whether the hydration requirement involved catalysis directly, or if it facilitated transport/diffusion of substrate and product, the impact of contamination density was tested. In this study, the neat

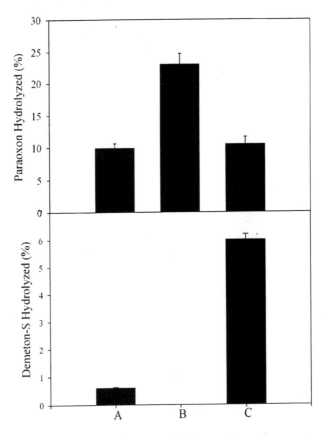

Figure 3. Hydrolysis over 24 hrs on dry surfaces.

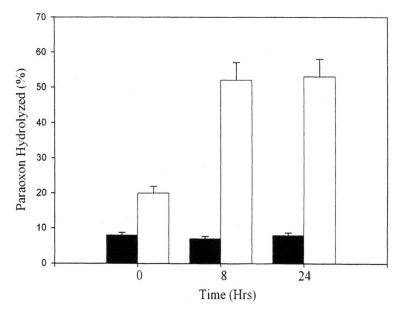

Figure 4. Recovery of activity by surface rehydration. Using the inactive surfaces (> 72 hrs drying), rehydration was performed by submerging the painted surface in water or buffer, followed by air drying. The black bars are the control panels (no rehydration) and the white bars represent the experimental panels.

OP compound was applied either as relatively large droplets (5 x 2 µl), or multiple smaller droplets (50 x 0.2 µl) or as a smear across the surface. In all cases, the same amount of OP was applied to surfaces that had been dried for greater than 72 hrs (Figure 5). As can been seen, even in the absence of re-hydration, the contamination density significantly impacted the enzyme-based reactivity. This further suggested that there may be adequate water remaining in these "off-the-shelf" paints for hydrolysis and the limiting factor is movement through the surface or coating for the OP substrate to contact the catalytic molecules. This can be addressed by the development of coatings with better hydrated characteristics and those with increased surface availability of the additive catalyst using increased porosity or ballooning techniques. However, these studies clearly show a role for OPDtox in off-the-shelf paints, which make these bioactive coatings more broadly applicable than specialty coatings. So, although the surfaces may lose their reactivity as they dry, the catalytic capacity remained active for months (even years), and they can be reactivated in the presence of adequate water to drive the reactions and/or facilitate mass transport of substrate to active site.

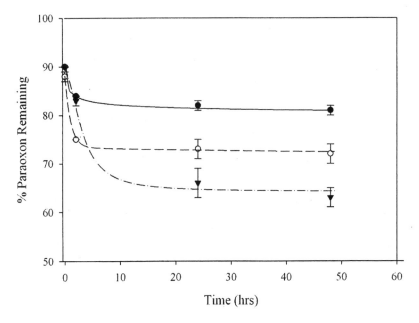

Figure 5. The effect of contamination density on surface reactivity. Neat paraoxon was applied at a contamination density of 5 x 2 µl (●), 20 x 0.5 µl (○) or as a smear across the surface (▼).

Summary

Enzyme catalysts offer substantial advantages over chemical catalysts in a traditional industrial environment. They are non-caustic, biodegradable, functional under mild conditions, provide benign alternatives to existing processes, and are not associated with the production of hazardous by-products or secondary waste. The current technology for OP decontamination involves the application of caustic solutions, foams, intensive heat and carbon dioxide for sustained periods. The biocatalytic coatings described here bring a new technology to bear on this problem, offering an opportunity for OP decontamination without the environmental impact. Finally, with the rapidly increasing database of natural enzyme diversity and recombinant DNA technologies, the protein modification tools exist to allow OP biocatalytic coatings to be tailored to specific applications. By incorporating other biomolecules, the application of enzyme-based additives for paints and coatings can be expanded into areas such as self-cleaning surfaces, mold-inhibiting surfaces, deodorizing surfaces, textile coatings, and catalytic coatings for waste stream decontamination.

[1]Abbreviations: tabun, O-methyl dimethylamidophosphorylcyanide; sarin, isopropyl methylphosphonofluoridate; soman, pinacolyl methylphosphonofluoridate; cyclosarin, cyclohexyl methylphosphonofluoridate; VX, O-ethyl S-diisopropylaminomethyl methylphosphonothioate; R-VX or VR, O-isobutyl S-(2-diethylamino)-methylphosphonothioate; Demeton-S, O,O-diethyl S-2-ethylthioethyl phosphorothioate; DFP, diisopropylfluorophosphate; Paraoxon, diethyl p-nitrophenyl phosphate

References

1. Gallo, M. A., Lawryk, N. J. In: The Handbook of Pesticide Toxicology; Editors. Hayes, W. J. Jr., Laws, E. R; Academic Press, San Diego, CA **1991**, pp. 920-925.

2. Mulchandani, A., Pan, S.T., Chen, W. *Biotechnol. Progr.* **1999**, *5*, 130-134.

3. LeJeune, K.E., Dravis, B.C., Yang, F., Hetro, A.D., Doctor, B.P., Russell, A.J. *Ann. NY Acad. Sci.* **1998**.

4. Tuovinen, K. *et al.*, *Fundam. Appl. Toxicol.* **1994**, *23*, 578-584.

5. Pravecek, L. M.S. Thesis. U.S. Naval Postgraduate School. Monterey, CA **2004**.

6. Fox, M. A., *Acc. Chem. Res.* **1983**, *16*, 314-321.

7. Yang, Y.-C. *et al.*, *J. Am. Chem. Soc.* **1990**, *112*, 6621-6627.

8. Gold, R.S., Wales, M.E., Wild, J.R. In: Enzymes in Action: Green Solutions for Chemical Problems; Editors: Zwanenburg, B., Mikoajczyk, M., Kiebasiski, P. NATO Science Series. Kluwer Academic Publishers. **2000**, pp.263-286

9. Russell, A.J., Berberich, J.A., Drevon, G.F., Koepsel, R.R. *Annu. Rev. Biomed. Eng.* **2003**, *5*, 1-27

10. Lai, K., Dave, K.I., Wild, J.R. *J.Biol.Chem.* **1994**, *269*, 16579-16584.

11. di Sioudi, B.D., Grimsley, J.K., Lai, K., and Wild, J.R. 1999. Modification of near active site residues in organophosphorus hydrolase reduces metal stoichiometry and alters substrate specificity. *Biochemistry* 38:2866-2872.

12. McDaniel, C. S., McDaniel, J., Wales, M.E., Wild, J.R. *Prog. Org. Coat.*, **2006**. In Press.

13. Simonian, A.L., Grimsley, J.K., Flounders, A.W., Schoeniger, J.S., Cheng, T-C., DeFrank, J.J., Wild, J.R. *Anal Chim Acta.* **2001**, *442*, 15-23.

Chapter 13

The Influence of Geometry on Superhydrophobicity

Oskar Werner[*] and Lars Wågberg

KTH, School of Chemistry, Department of Fibre and Polymer Technology, Division of Fibre Technology, Teknikringen 58, 10044 Stockholm, Sweden
[*]Corresponding author: email: owerner@polymer.kth.se

Super-hydrophobic surface properties may arise from the interplay between an intrinsic, relatively high contact angle of the solid surface involved, and the geometric features of the solid surface. In the present work this relationship was investigated, for a range of different surface geometries, making use of a theoretical model based on surface free energy minimisation. As a rule, the free energy minima (and maxima) occur when the Laplace and Young conditions are simultaneously fulfilled. Special efforts have been devoted to investigating the free energy barriers that are present between the Cassie-Baxter (heterogeneous wetting) and Wenzel (homogeneous wetting) modes of wetting. Along with the above scheme a new experimental method for characterising the wetting of structured surfaces has been developed. Sessile drops of 3 % (by weight) agarose solution were immobilised on test surfaces. The drops could be removed after solidification and by using confocal microscopy and image processing it was possible to characterise the interface between the droplet and the solid surface. This analysis also made it possible to determine wetting mode of the droplet, and to estimate the wet surface area and the local contact angles beneath the drop.

Background

In recent years there has been a growing interest in super hydrophobic surfaces. In this paper both a theoretical model describing a drop on surface system in heterogeneous wetting mode and a method for characterising the liquid — vapour interface beneath a drop in heterogeneous wetting mode is presented. The precise definition of the term "superhydrophobic" is debated but it is fairly commonly to apply to a condition where a water drop, applied to such a surface, should has a fairly high apparent contact angle (approximately 150°) and a very low sliding angle. In an absolute majority of cases this means that the drop is in the heterogeneous (Cassie – Baxter) wetting regime i.e., the area under the drop is not completely wetted. Due to its importance and complexity, the interest in the phenomenon of heterogeneous wetting has increased and several previous works have dealt with its theory from various perspectives. (1-11) Even though the phenomenon can in principle be defined by the fulfilment of the Young and Laplace conditions, the number of different states that need to be examined are often vast – even for a rather simple surface geometry. Alternatively a drop on a surface can be described by considering the system as a whole. The difference between the two viewpoints can be regarded as the difference between considering the free energy balance at every point on the liquid-vapour interface (the Laplace Condition) and at every point along the three phase contact lines (the Young condition) or to see the drop as a system subject to certain constraints and while taking account of the various free energy contributions. An obvious drawback of the latter approach is that it means that certain approximations have to be introduced. On the other hand, this approach yields information regarding the stability of the wetting modes and the height of the free energy barrier between them: this is the approach used been in the present work.

In testing heterogeneous wetting theories determination of the liquid-vapour interface beneath the drop is of importance though obviously difficult. One way to solve this problem is to preserve the geometry of the interface and, remove the drop to allow examination of it. Initial experiments using a novel method to accomplish this are presented in this work.

Theory

Local view

For a drop in a stable state of heterogeneous wetting state there has to be pressure balance at each point on the liquid – vapour interface. This balance is governed by the Laplace expression,

$$\Delta P = \gamma \left(\frac{1}{R_1} + \frac{1}{R_2} \right), \qquad (1)$$

where ΔP is the pressure difference across the liquid-vapour interfaces, γ is the interfacial free energy, R_1 and R_2 are the radii of curvature of the liquid-vapour interface.

At the three phase contact lines the local contact angle is governed by the Young equation,

$$\cos\theta = \frac{\gamma_{sv} - \gamma_{sl}}{\gamma_{lv}}, \qquad (2)$$

where, γ_{sv}, γ_{sl} and γ_{lv} are the solid–vapour, sold–liquid and liquid–vapour interfacial energies respectively. If gravity effects are set aside, the pressure will be uniform in the drop. If the surrounding pressure also is uniform (not necessarily so in the case of trapped bubbles) ΔP and likewise the total curvature ($1/R_1 + 1/R_2$) will be the same over the entire liquid–vapour interface. However, if the gravity effects are significant the pressure will, of course, vary inside the drop. To use the above equations (for every point) to investigate whether a drop will be in heterogeneous or homogenous wetting mode, and to obtain their respective degree of stability, usually entails the investigation of a vast number of states.

The global view

Another possibility is to consider the drop as a system subject to certain constraints, moving in a particular state-space and invoke the different free energy contributions. An obvious drawback of this approach is that certain approximations must be introduced but on the other hand this approach will provide additional information concerning the system, in particularly the stability of the wetting modes, and the height of the free energy barrier between them. A model of this kind is presented in the next section. Patankar.(5), showed with analytical energy analysis that any existing state between the Wenzel and Cassie minima would indeed be of higher energy and represent an energy barrier. In the work presented here the focus has been on these energy barriers.

A new wetting model using the global view

A global approach is taken by considering the overall excess free energy of the whole system comprising a droplet resting on a solid surface. The surface free energies are invoked as well as the potential energy of the water in the droplet in the gravitational field. In this way, every point in the *state-space*, in which each dimension represents one of the system variables, corresponds to a particular free energy value. This energy surface can be used in determining the thermodynamically stable and metastable states, and also for determining the depth of the corresponding minima and the heights of the barriers between them.

The model uses two variables: the apparent contact angle, θ, and the penetration depth of the liquid into the base structure, z. Hence, the state-space has two dimensions. The geometrical surface parameters, f (i.e. the wetted fraction of the projected area) and r (i.e. the real wetted area per projected wetted area), are considered to be functions of z (Figure 1). Thus $f(z)$ and $r(z)$, together with the interfacial free energies of the *solid–liquid* and *solid–vapour* interfaces, γ_{sl} and γ_{sv}, are used as the sole characteristics of the solid surface.

The thermodynamic scheme of both the Young and the Cassie-Baxter equations describing the droplet-on-surface problem can be traced back to minimisation of the free energy (hereafter denoted E) with respect to θ. According to the Cassie-Baxter approach, (1) the *average* free energy per unit area beneath the droplet is considered when calculating the apparent contact angle, θ, the result being the Cassie-Baxter equation,

$$\cos \theta = \frac{fr(\gamma_{sv} - \gamma_{sl})}{\gamma_{lv}} + f - 1 = fr \cos \theta_{intr} + f - 1, \tag{3}$$

where the intrinsic contact angle, θ_{intr} is to be distinguished from the apparent contact angle, θ. Along the z-axis, the fulfilled Young condition (using the zero hydrostatic pressure approximation, resulting in a planar liquid-vapour interface beneath the droplet) and knowledge of the geometry of the surface (parallel cylinders in the original work) (1) are used to calculate the penetration depth of the liquid into the structure. In the case of $r = 1$, Eqn. (3) becomes $\cos \theta = f \cos \theta_{intr} + f - 1$, and in the case of complete wetting it reduces to the Wenzel equation (2),

$$\cos \theta = r \cos \theta_{intr}. \tag{4}$$

Concerning the hydrostatic and capillary pressures, one might invoke the pressure-induced curvature of the liquid–vapour interface beneath the droplet. For a global model, however, this is not very convenient, since the curvature will be a quite complex function of the spatial co-ordinates, for even a rather simple surface topography.

In the present model the assumption that the liquid—vapour interface beneath the droplet is flat is retained, in the sense that the extension of the liquid-vapour interface is estimated as if it were actually planar. (The planar approximation is good for pore sizes up to approximately one tenth of the droplet radius; this can be checked using the chord theorem.) Instead, the changes in the droplet gravitational energy are considered explicitly. Hence, when the interface propagates in the z direction the gravitational energy decreases. To determine the equilibrium penetration, z, this decrease is balanced against the net interfacial free energy change due to the change in interfacial area. The situation encountered here is in principle the same as that found in the case of capillary rise. The effect of gravity can be accounted for either by considering the Laplace pressure drop across the liquid–vapour interface or in terms of the potential of the rising liquid in the gravitational field. (*13*)

A droplet resting on a solid surface is treated as a system located somewhere in a 2D state-space specified by the two variables, θ and z. For each system, every state (θ, z) corresponds to a certain E value. It is then possible to determine the path along which the system must move in order to reach a state of lower E. A plot of such a state-space is shown in Figure 2. The separate contributions to E are shown in Eqns. (6-9) below. The first term, describing the gravity contribution Eqn. (6), indicates how the free energy will decrease when the droplet's centre of mass declines with increasing values of z. For example, if $f(z)$ is 0.5 for all values of z, the centre of mass will decline 1 μm when z increases 2 μm. Eqns. (7–9) imply that the interfacial free energy contributions are proportional to the area of each kind of interface. The constant in Eqn. (9) represents the total interfacial energy of the *solid-vapour* interface of the non-wetted sample. This constant has been set to zero in this model; however, this is not a restriction, since only the change in area is of significance. In mathematical formulation;

$$variables \quad : \quad \theta, z$$

$$constants \quad : \quad \gamma_{lv}, \gamma_{sv}, \gamma_{sl}, V, g, \rho$$

$$known\ functions \quad : \quad R(\theta),\ r(z),\ f(z)$$

where θ is the apparent equilibrium contact angle (degrees), z is the liquid's penetration depth into the surface structure (m), γ_{ij} is the interfacial free energy (Jm^{-2}), V is the volume(m^3), ρ is the mass density (kgm^{-3}), g is the gravitational acceleration (9.82 ms^{-2}) R is the drop radius (m), $f(z)$ is the wetted fraction of the projected area (dimensionless), $r(z)$ is the real wetted area per projected wetted area (dimensionless), and E is the overall free energy(J)

$$E = E_{lv} + E_{sl} + E_{sv} - E_{gravity} \tag{5}$$

$$E_{gravity} = V_{\rho gz}\left(1 - \frac{\int_0^z f(x)dx}{z}\right) \tag{6}$$

$$E_{lv} = \gamma_{lv}S_{lv} = 2\pi R^2\gamma_{lv}(1 - \cos\theta) + \\ + (1 - f(z))\gamma_{lv}\pi R^2 \sin^2\theta \tag{7}$$

$$E_{sl} = \gamma_{sl}S_{sl} = f(z)r(z)\pi R^2\gamma_{sl}sin^2\theta \tag{8}$$

$$E_{sv} = \gamma_{sv}S_{sv} = const. - f(z)r(z)\pi R^2\gamma_{sv}sin^2\theta \tag{9}$$

i.e.

$$E(z,\theta) = \pi R^2 (2\gamma_{lv}(1-\cos\theta) + \sin^2\theta((1 - f(z))\gamma_{lv} +$$
$$+ f(z)r(z)(\gamma_{sv} - \gamma_{sl}))) - V\rho g z \left(1 - \frac{\int_0^z f(x)dx}{z}\right) + const. \tag{10}$$

where the functions $r(z)$ and $f(z)$ are specific for each particular surface topography. A schematic explanation is found in Figure 1. If the influence of gravity is neglected, i.e. $E_{gravity} = 0$, z is set to a constant value and $\theta_i = (\gamma_{sv} - \gamma_{sl})/\gamma_{lv}$, then expression (10) reduces to $E(\theta) = \pi R^2(\theta)(\sin^2\theta (\gamma_{sl} - \gamma_{sv}) + 2\gamma_{lv}(1 - \cos\theta))$ + $const$, and $\delta E / \delta\theta = 0$ will give Eqn. (1).

Implicit in expression (10) is the assumption that the amount of liquid penetrating the surface structure is small compared to the droplet volume. (Taking account of the decrease in volume above the droplet base and the ensuing decrease in R and E_{lv} corresponds to considering the Laplace pressure of the spherical droplet surface in the Young-Laplace approach.) The present model accounts for gravity to only a limited extent. The pressure differences inside the droplet and the non-spherical droplet shape they cause have been left out of consideration.

In summary; Eqn. (6) describe how the potential energy vary with z, taking into account that the altitude of the centre of mass is not directly proportional to z. Eqns. (7-9) are to a large extent expressing the geometry of a spherical cap of which the spherical surface is multiplied with the surface tension of water. The base area is weighted, according to the principles in Figure 1 to obtain the wet are and the liquid-vapour area beneath the drop each is multiplied the respective surface energy. Eqn.(10) is Eqns (6-9) in Eqn (5).

Simulated drop on surface systems

A computer tool was coded in MATLAB in order to test the applicability of the main expression presented in Eqn. (10). For a structure of given parameters and surface functions the program calculates a matrix consisting of $E(\theta, z)$ values, according to Eqn. (10), for all relevant values of (θ, z). This matrix, the state-space, maps all the possible system states. The global minimum represents a stable state, while the local minima represent metastable states.

The program-generated matrix containing the energy values of all (θ, z) values is displayed as a free energy-surface (Figure 2). Bearing in mind that the system will go from higher to lower free energy states, this plot gives a general idea as to how the system will behave. A droplet-on-surface system will move from its initial state towards states of lower free energy until it comes to rest at a free energy minimum, which could be either local or global. For a system at rest to start moving towards higher free energy states, external disturbances such as vibrations are needed.

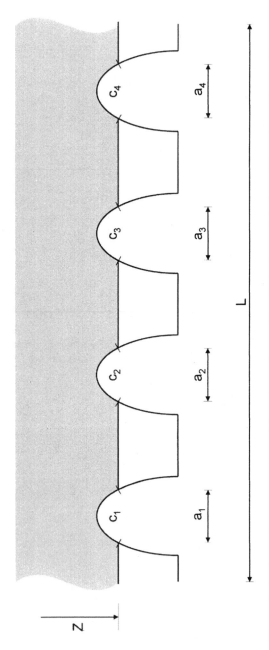

Figure 1. A schematic representation of the z-dependence of f and r in the model with a 2D analogy. The liquid-vapour interface beneath the droplet can be approximated as being flat if the structure is small in scale compared to the droplet radius. Hence, the penetration of the water into the structure can be described with one sole measure, z. In this figure, where $\Sigma_i c_i$ is the real wetted sample distance and $\Sigma_i a_i$ is the projected wetted sample distance, f_{2D} and r_{2D} are defined as

$$f_{2D} = \sum_i \frac{a_i}{L}, \quad r_{2D} = \frac{\sum_i c_i}{\sum_i a_i}$$

Both a_i and c_i, and hence r_{2D} and f_{2D}, are dependent on z. While the variable, z, in the model Eqn. (5-10) describes the water penetration, the functions f and r describe the geometry of the surface itself. (Reproduced from reference 14. Copyright 2005 American Chemical Society.)

To describe more accurately where these minima are located, the lowest free energy, and the corresponding θ, is found for every value of z (Figure 3). These results are obtained by minimising $E(\theta,z)$ for every z with respect to θ (Eqn. 10). The plot of this path in the state-space is of particular interest, as it shows how the free energy of the system will vary as z increases. A more thorough description of this model is found in Werner et al. *(14)*, where the model is also tested against experimental results found in the literature. It should also be mentioned that during the early development of the present theoretical model describing wetting of super-hydrophobic surfaces *(15)* a similar but still significantly different model was simultaneously developed by Jopp et al. *(9)*

Stability of wetting modes

The investigation of a droplet on a solid surface is not completed when just the minima have been located, it is almost equally important to determine the barriers between the minima. If a droplet is in a metastable state and the barrier is low, even a slight vibration could suffice to push the droplet into the stable state. In this study "barrier height" by itself is understood as the relative free energy difference between the local free energy minimum in question and the point with the highest free energy (usually a saddle point in the state-space) the system must pass to reach the stable minimum. The program locates the local maxima. The "barrier height from X" should be understood as the barrier height that must be crossed in order to leave state X.

A way to explicitly describe how the height of the barrier from both directions changes with some variable, here the intrinsic contact angle, is depicted in Figure 4 in which the free energy for complete wetting, for the Cassie–Baxter minimum (when existing), and for the maximum is plotted against θ_{intr} for a particular surface geometry and droplet volume. The maximum should be understood as a maximum in terms of $E(z)$, as in Figure 3, which represents the threshold between two (meta)stable states - if two such exist. The Plot For a given geometry, the barrier from Cassie–Baxter can be obtained from the diagram by reading the vertical distance from the Cassie–Baxter curve to the maximum curve at the intrinsic contact angle of interest.

As seen in Figure 5, for a surface with small, closely spaced cavities, both the Wenzel and the Cassie–Baxter minima can be of the same depth; for high enough values of θ_{intr}, the Wenzel minimum will even cease to exist. Thus not only in transitions from Cassie–Baxter to Wenzel mode are of interest, but also the reverse direction from Wenzel to Cassie–Baxter mode. When investigating such transitions, it will likely be very important to study local effects, such as the spontaneous formation of vapour bubbles and whether they will have the ability to spread, merge, and form a continuous liquid–vapour interface.

258

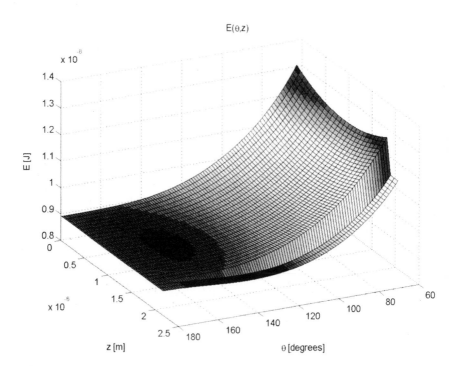

Figure 2. A 3D plot of the free energy E*(θ,z) of a 4 μL-droplet on an idealised surface partly covered with hemispheres with radii of 2μm. The step at* z = 2μm *corresponds to the transition to complete wetting, whereby the area of the high-energy* liquid-vapour *interface decreases. A spherical droplet placed on top of the surface is assumed to be in the initial state, (180°, 0 μm). Since this is not an equilibrium state, the system will shift towards a lower free energy and settle at the free energy minimum (150°, 1 μm) (see also Figure 3). If the droplet is disturbed, for example, by being mechanically depressed, it could be forced past the free energy threshold settling instead at the global minimum at (135°, 2 μm), which corresponds to homogenous wetting for the system in question. (Reproduced from reference 14. Copyright 2005 American Chemical Society.)*

Experimental

Investigation of liquid–vapour interface beneath the droplet

The liquid–vapour interface beneath a drop in heterogeneous wetting mode is of considerable interest when investigating a given superhydrophobic surface. It is also rather difficult to characterise quantitatively. (17) In this study, the development of a method for investigating the general appearance of this interface is presented. The underlying concept for this method involves the gelation, removal and 3D characterisation of the droplet. The first step is to place a drop of water-based agarose solution, heated to 70°C on the surface to be examined. At this temperature the agarose solution behaves essentially as water would do at the same temperature (cf the monomeric liquid used in Krupenkin et al. (17)). As the drop cools to room temperature the agarose molecules become interlocked via hydrogen bonding and the drop solidifies to form a stable gel. The water content and shape of the drop has been observed to remain essentially unchanged throughout this process. The gelled drop can then be removed and its underside investigated (Figure 6). In the present work a confocal microscope was used to make an actual height map of the solid drop. Together with knowledge of the properties of the studied base-surface this height map allows the shape of the liquid–vapour interface to be determined in the event of heterogeneous wetting.

Preparation of Model Surfaces

Printing plates etched with a screen pattern were used as a model system in this study. Printing plates of type Nyloprint WF-M were provided by BASF, who also conducted the optochemical etching. The plates used had resolutions of 54 and 80 lines per centimetre and tonal values of 2, 5, 10, 20, and 50% (Figure 7). The composition of the plates by weight was as follows a) a mixture of poly(vinylalcohol) and poly(vinylalcohol-co-polyethyleneglycol), 60 %, b), mixture of monoacrylate and diacrylate, 38 %, and c) photoinitiator and stabiliser dye, 2%.

To render the plates more hydrophobic, they were treated with octadecyltrichlorosilane (OTS) by means of vapour deposition. This was done by placing the plates upside down over a petri dish containing the hydrophobic agent. The distance between the plate and the liquid-phase OTS was 1 cm. To prevent the OTS from reacting with the oxygen of the air, this procedure was conducted under a nitrogen atmosphere. Before the treatment the plates were carefully cleaned using an excess of deionised water and then dried. Goniometry revealed the contact angle of water to be 85 ± 3° on a smooth OTS-treated printing plate, compared to 45 ± 3° on a smooth untreated plate.

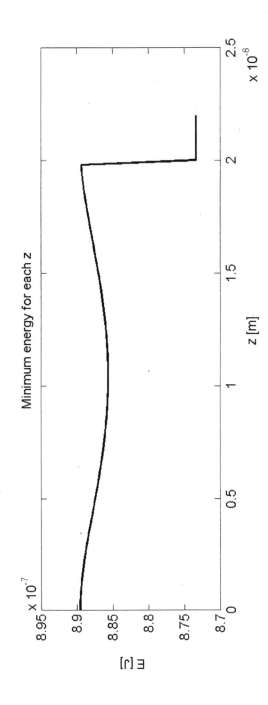

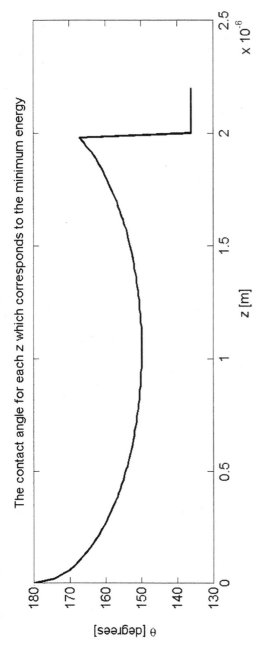

Figure 3. Graph showing the $E(\theta_{eq}, z)$ *function for the hemisphere-covered surface described in Figure 2. For every value of z, the minimal free energy and corresponding value of θ is shown. For this surface there is a local minimum at (150°, 1 μm) and a global minimum at (135°, 2 μm) for both the free energy and contact angle. The straight horizontal line to the far right is included simply to make the value at z = z_{max} easier to read.*

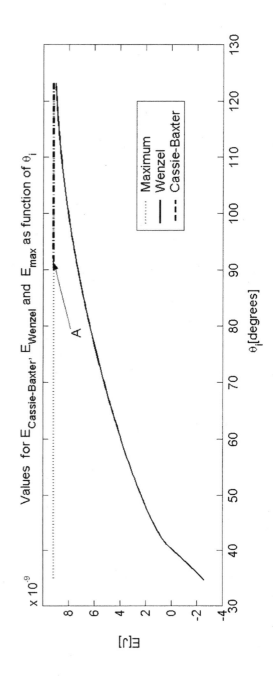

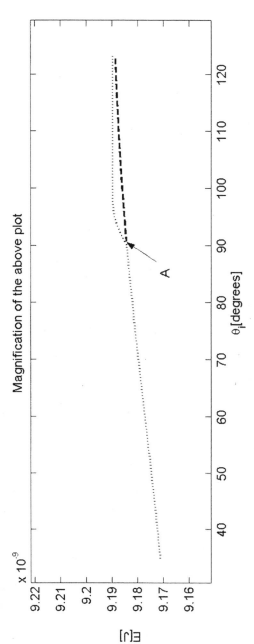

Figure 4. *The free energies for complete wetting, for the CassieBaxter minimum (if present), and for the local maximum are plotted against θ_{intr} for a particular surface geometry (here the pillar surface presented by Bico et al. (16) and droplet volume $V = 4.2$ nl). The maxima should be understood as the points diagram of the type found in Figure 3 with no higher values in adjacency, in the case depicted in Figure 3 two maxima are present since also the end value will be counted as maximum with this definition. In the examples in Figures 4 and 5 only one maximum is present though. For each θ_{intr} the height of the thresholds can be read from the diagram, as can the number of minima and their positions. At point A in the figure, the maximum energy no longer corresponds to that of the droplet resting on top of the pillars. At this point the metastable Cassie–Baxter minimum is found (this minimum is nonexistent below this point). The threshold the system must pass to get from the Cassie–Baxter minimum to the much deeper Wenzel minimum should be read as the vertical distance between the respective curves. In comparison, the potential energy of a 4.2 nl-water droplet elevated 1 dm is 4.1 nJ.*

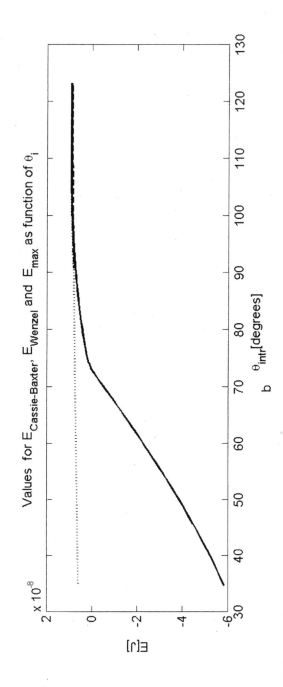

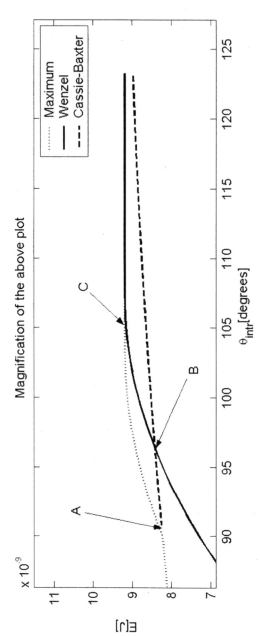

Figure 5. The free energies for complete wetting, for the Cassie-Baxter minimum (if present), and for the threshold are plotted against θ_{intr} for a particular surface geometry and droplet volume (here a simulated surface covered with cavities and $V = 4.2$ nl). As can be seen, the Wenzel minimum is the only one existing until A. Between A and B the Cassie–Baxter minimum is shallower than the Wenzel minimum is. The depths at each θ_{intr} can be read as the vertical distance from the respective state to the maximum value. At B the two minima have the same free energy, but still have a considerable barrier between them. At even higher contact angles the threshold from Wenzel to Cassie–Baxter shrinks until it ceases to exist at C.

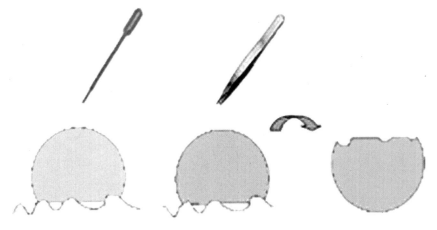

Figure 6. A drop of 70°C agarose solution was applied to the surface with a pipette. Being comparatively small, the drop quickly cooled down to temperatures below the gel point forming a stable gel. The now stable gel drop could be lifted from the surface and its underside examined.

Preparation of Agarose Drops

A 3% solution of agarose, type D-5 (Hispanagar, Burgos, Spain), with a molecular weight of 200 000 according to the supplier, was prepared. The agarose was stirred at 70 °C for 2h in deionised water. It is crucial not to heat the solution for too long, or hydrolysis could preclude the formation of a stable gel. To make it possible to use a confocal microscope in fluorescence mode, 10 μg/ml of the fluorescent agent acridine orange (Fluka), was added to the solution. With the aid of a pipette 10 μl drops (giving drop diameters in the mm range) were applied on the surfaces where they cooled down and gelled in a few seconds. The drops were then carefully removed and placed upside down on slides which were fitted into the confocal microscope. The gelled drops kept their shape for about half an hour at room atmosphere and for more than a week in a cool and humid environment.

Image Acquisition

The confocal microscope, in this study a BioRad with a Radience 2000 argon krypton laser, uses the principle of a moving focal plane. The procedure can be likened to taking a number of pictures using an extremely shallow depth of field: in each picture only a thin section located at the focal plane will be shown. In this study 512×512 pixel images of 600×600 μm^2 sample areas were collected and the information from them was used to quantify the 3D

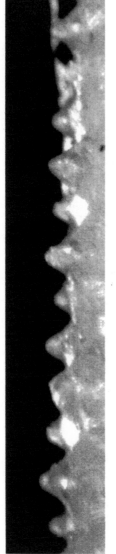

Figure 7. A cross-section of one of the test surfaces (54 lpc, 2% tonal value). The lines per cm values (lpc) for the screen patterns used in this work were 54 lpc and 80 lpc, giving distances between adjacent peaks of 209.05 µm and 141.25 µm, respectively.

features of the sample. The resolution in the z-direction was 2 μm. This information can be superimposed to produce a result resembling an ordinary image complete with both illuminated areas and shadows. The images can also be used to create a height map of the sample.

Initial Experimental Results

Characterisation of Agarose Solution

Agarose molecules are highly hydrophilic and therefore not expected to change their surface tension to any greater extent. To investigate this matter, DuNouy ring surface tension measurements were conducted for the agarose solution. The surface tension of water at 70°C is 65 mN/m (73 mN/m at room temperature). The surface tension of the agarose solution was measured both with and without the addition of the fluorescent agent revealing that the fluorescent agent had very little influence on the results obtained. In both situations the solution at 70°C had an initial surface tension of 54 mN/m, which decreased only little and slowly with time.

Wetting of Periodically Structured Polymer plates

The drops taken from the printing plates indicated that the agarose solution had wetted the surface completely, except where vapour bubbles were located. As seen in Figure 8 both the solid–liquid and the solid–vapour interfaces are distinctive.

Height Maps

Scanning many thin sections of a sample produces a 3D array of scanning points. Each pixel position in the x–y plane corresponds to a set of scanning points along the z-axis. The value of z corresponding to the point with the highest light intensity is assigned to this pixel position, so as to create a height map of the agarose sample (Figure9 a)). Figure 9 b) shows a profile of the agarose surface, taken along the transect indicated by the black line in Figure 9 a). It displays three distinctive dips. Those to the left and to the right correspond to the former liquid–solid interface. Since the liquid agarose solution was here in direct contact with the solid surface the gelled drop in these areas is a direct cast of the substrate. In contrast the dip in the middle represents a cast, not of the solid surface, but of an entrapped air bubble. The profile, and the height map, can be used to derive quantitative information on the liquid–vapour interface beneath the drop.

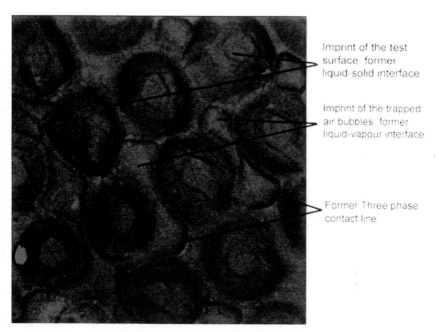

Imprint of the test surface, former liquid-solid interface

Imprint of the trapped air bubbles, former liquid-vapour interface

Former Three phase contact line

Figure 8. The former phase interfaces ("former" here refers to the condition in when the drop was fluid and resting on the surface) can be identified on a gelled and removed drop. In this figure the contour of the former three-phase contact line is clearly visible on the superposed images, as is the shape of the former liquid–vapour interface with its nodes and loops. The large round features are imprints of spikes like those depicted in Figure 7. (See page 4 of color insert.)

Discussion

Quantification of Energy barriers

Many of examples of the superhydrophobic wetting state being metastable are found in nature and many are also described the literature. For example, a drop carefully placed on the fur-like surface of the leafs of *Senecio cineraria* (dusty miller) will be almost spherical while a drop dropped onto it from one metre will wet the leaf. Bico et al. *(16)* have reported that the contact angle of a drop placed on a superhydrophobic pillar surface decreased dramatically and irreversibly when a slight pressure was applied. Quite recently Cheng et al. *(18)* condensed water onto a lotus leaf achieving non-superhydrophobic behaviour. In the of *Senecio cineraria*, the surface is likely deformed during this transition, due to the softness of the surface hairs. To get an appropriate model for this transition it would then be necessary to include the mechanical properties of the surface. This has already been suggested in the case of Lady's mantle. *(7)*

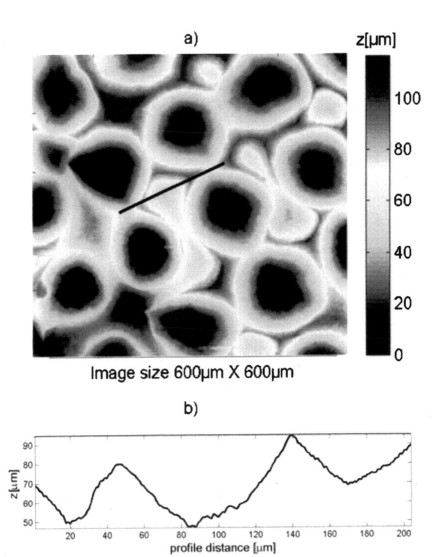

Figure 9. This figure presents, a) a height map of the underside of a drop
(See page 4 of color insert.), and b) a profile across a bubble, taken along the
black transect indicated in a). Using this method it is possible to obtain
quantitative information concerning the geometry of the liquid–vapour
interface.

In the case described by Bico et al. *(16)*, the surface is rigid but the pillars are sparsely placed and the energy barrier is so low that that a slight pressure is sufficient to force the system over the barrier. In case described by Cheng et al *(18)* the water is condensed on the surface. This experiment can be compared with a drop on surface system starting at $(0°, z_{max})$ instead of at $(180°, 0 \ \mu m)$. A stable or metastable Wenzel state exists even with the sacred lotus: however, the barrier is high and the leaves work as effective self cleaning surfaces.

A way to measure the free energy barriers quantitatively could be to let drops fall from different heights onto the tested surfaces and then measure the contact angles. This way it would be possible to investigate how the critical height, or critical kinetic energy at impact, would depend on the geometric properties of a surface.

Further development

The main obstacles for this method lie in the characterisation and it is on this aspect future development will be focused. Since the drop, even in its gel state, mainly consists of water it shrinks somewhat when exposed to the laser of the confocal microscope. This shrinkage is not great enough, however, to seriously distort the features of the surface during measurement. The measured surface will, however, retreat very slowly though not insignificantly, due to the total shrinkage of the whole drop. Since the scanning is done from the top, the 3D image will be slightly stretched in the z-direction. One way to deal with this would be to measure the speed of shrinkage and then compensate for it in the final image.

Further testing of relationship between the gelation and wetting speed of agarose solution and how well the agarose gel withstands removal also has to be conducted, even though initial results look promising in this matter

Conclusions

The essentially thermodynamic model resulting in the Cassie–Baxter equation has been generalised into a model that also covers the balance between the free energy change due to gravity (potential energy) and the free energy change arising from changes in area of the liquid–vapour, solid–vapour, and solid–liquid interfaces. For a given surface topography and a given intrinsic contact angle, free energy as a function of the contact angle and the liquid's penetration of the surface structure, $E(\theta, z)$, yielded more information regarding the surface's water-repellent properties than did formulae such as Wenzel's or Cassie's, especially concerning the stability of the wetting modes.

The free energy barriers existing between the wetting modes were also studied. This showed that it is insufficient to know where the minima are located and which is the lowest; it is also of interest to identify the energy barriers existing between the minima themselves and between the minima and the droplet's starting position in the state-space. Falling-drop experiments are proposed as a method for investigating these barriers.

According to the present global model, surfaces with nonstable Wenzel wetting mode exist. To further study the nonstable Wenzel mode it will likely be necessary to investigate local effects, such as the spontaneous formation of vapour bubbles and whether they will have the ability to spread, merge and form a continuous liquid–vapour interface.

Apart from the theoretical model a promising new experimental technique for characterising the liquid–vapour interface by solidifying agarose solution droplets were developed and described. Due to the waterlike properties of the agarose solution it gives, unlike direct casting of a surface, information of how the liquid—vapour interface beneath a droplet in heterogeneous wetting mode.

Acknowledgements

Professor J. C. Eriksson is acknowledged for discussions on the theories of wetting. Bert Petterson is acknowledged for help with the agarose chemistry. The authors also wish to thank Anni Hagberg at STFI for help with the confocal microscopy, and Dr Uwe Stebani at BASF for supplying the patterned printing plates. Finally, Vinnova and FPIRC is acknowledged for financing the work of O. Werner. And, Bo Rydins Stiftelse för Vetenskaplig Forskning, for financing the participation in Smart Coatings 2006.

References

1. Cassie, A.; Baxter, S. *Trans Faraday Society* **1944**, *40*, 546-551.
2. Johnson, R.; Dettre, R. Surface and Colloid Science. In, Vol. 2; Science, E. M. C., Ed.; John Wiley and Sons, Inc: New York, 1969; Chapter Wettability and Contact Angles, pages 85-153.
3. Oliver, J.; Huh, C.; Mason, S. *Colloids and Surfaces* **1980**, *1*, 70-104.
4. Kijlstra, J.; Reihs, K.; Klamt, A. *Colloids and Surfaces* **2002**, *206*, 521-529.
5. Patankar, N. *Langmuir* **2003**, *19*, 1249-1253.
6. Marmur, A. *Langmuir* **2003**, *19*, 8343-8348.
7. Otten, A.; Herminghaus, S. *Langmuir* **2004**, *20*, 2405-2408.
8. Zhang, J.; Kwok, D. *Journal of Colloid and Interface Science 282* **2005**, *282*, 434438.
9. Jopp, J.; Grüll, H.; Yerushalmi-Rosen, R. *Langmuir* **2004**, *20*, 10015-10019.
10. Pal, S.; Weiss, H.; Keller, H.; Müller-Plathe, F. *Langmuir* **2005**, *21*, 3699-3709.

11. Dupuis, A.; Yeomans, J. M. *Langmuir* **2005**, *21*, 2624-2629.
12. Wenzel, R. *Industrial and Engineering Chemical Research* **1936**, *28*, 988-994.
13. Henriksson, U.; Eriksson, J. C. *Journal of Chemical Education* **2004**, *81*, 150-154.
14. Werner, O.; Wågberg, L.; Lindström, T. *Langmuir* **2005**, *Vol. 21, No. 26*, 12235-12243
15. Werner, O. "Computer Modelling of the Influence of Surface Topography on Water Repellencyand a Study on Hydrophobic Paper Surfaces with Partly Controlled Roughness", Master's thesis, Linkoping University, 2003 http://www.ep.liu.se/exjobb/ifm/tff/2003/1168/exjobb.pdf.
16. Bico, J.; Marzolin, C.; Quere, D. *Europhysics Letters* **1999**, *47*, 220-226.
17. Krupenkin, T. N.; Taylor, J. A.; Schneider, T. M.; Yang, S. *Langmuir* **2004**, *20*, 3824-3827.
18. Cheng, Y.-T.; Rodak, D. E. *Applied Physics Letters* **2005**, *86*, 144101

Chapter 14

Intelligent Corrosion Protection by Conducting Polymers

Michael Rohwerder

Max Planck Institute für Eisenforschung, Max-Planck-Strasse 1, 40237 Düsseldorf, Germany

Use of intrinsically conducting polymer (ICP) coatings for corrosion protection has been discussed controversially for years. Quite a number of possible protection mechanisms are proposed, of which the possible passivation of the metal through the high potential of redox ICPs such as polypyrrole or polyaniline is maybe the most frequently stated. In this presentation the focus will be on the dynamic response mechanisms that such redox ICPs will show when a delamination front approaches them.

Of special interest is the possibility of intelligent self-healing by conducting polymers that show a true electrochemically driven on-demand release of inhibitor anions. The trigger for the release is the potential decrease that occurs during delamination at the polymer/metal interface. It will be shown that the efficacy of conducting polymers for corrosion protection very much depends on how they are applied and on the conditions of the corrosion experiment, i.e. depending on the exact conditions a conducting polymer may have excellent protection capability or may lead to a disastrously enhanced corrosive attack. This paper will discuss how to counteract the negative properties of the conducting polymer, so as to truly benefit from the positive ones.

Introduction

Organic coatings have a widespread application for corrosion protection of metals. In order to limit corrosion of the metal substrate, corrosion inhibiting pigments are added to the paints and other organic coatings applied on metallic surfaces. The most efficient pigments are those containing chromates (usually in form of strontium chromate), but because of their toxic and carcinogenic nature their use has to be progressively decreased. But nearly all powerful inhibitors may have detrimental effects on environment, when released in substantial amounts. Since in basically all pigments the release of inhibitors is based on leaching, coatings need to be highly pigmented to ensure a sufficient presence of inhibitors over years, and, of course, inhibitors are constantly released into the environment, even when they are not needed. Hence, novel approaches are desperately sought. Of special interest are of course concepts where no unwanted leaching will occur and release of active agents will take place only when necessary because of corrosive attack.

This is more true as the environmental issues become increasingly important – in Europe according to a new EC directive the chemical industry has to re-evaluate many of the approximately 100.000 chemicals on the market in respect to their toxicity. And in addition to that the requirements in industry become more and more demanding.

For instance, the trend in automotive industry goes for coil-coated steel sheet that can be formed and cut without loss of performance. This so-called "finish first, fabricate later" concept increases the value of steel sheet supplied by the steel industry and allows the automotive industry to cut one more production step. However, forming of non-elastic, plastic material (such as the metal substrate) unavoidably results in al least nanoscopic defects at the interface where slip bands appear [1], which have a direct negative impact on the coating performance.

At defects, sooner or later corrosion will occur and cause electrochemically driven delamination of the coating. Especially the so-called cathodic delamination can be extremely fast (see e.g. [2]).

To counteract the corrosive attack at the interface is the task of the pigments added to the organic coating. One class of pigments that are currently investigated by the corrosion science community are cation-exchange pigments containing divalent alkaline earth cations such as Mg^{2+}, Va^{2+} or Ca^{2+} or trivalent rare earth metal cations such as Ce^{3+} or Y^{3+}, some of which may have a promising potential for preventing cut edge corrosion. While these pigments do not leach when in contact with pure water, they release their cations when in contact with electrolyte (e.g. at the corroding defect). However, in many application the intact coating may come into contact with salt (e.g. near the sea or in industrial areas) and unwanted leaching may occur. Also, the efficacy at larger defects such as at cut edges still does not compare to that of chromate pigments.

Since about two decades the potential of conducting polymers for corrosion protection in coatings is topic of controversial discussion [see e.g.[3; 4]. Up to now, only little is known about how corrosion protection by conducting polymers might work. A number of different mechanisms are proposed, such as the so-called "ennobling mechanism", that is based on the assumption that conductive redox polymers such as polyaniline or polypyrrole, applied in their oxidized state, may act as an oxidizer, improving the oxide layer at the polymer/metal interface (see e.g. [5]) or even maintain the metal in small defects in the passive domain [6; 7]. Another proposed mechanism is that conductive polymers might shift the reaction site of oxygen reduction, the key reaction during delamination, from the metal/polymer interface into the polymer and thus smear out the produced radicals that destroy the adhesion over the full coating, thus significantly lowering their concentration [8; 9]. A detailed analysis of these mechanisms can be found in another paper [10].

The focus of this presentation is the possible ability of conductive redox polymers to release corrosion inhibitors on demand. Barisci et al. [11] for the first time pointed out that, that as a result of a galvanic coupling between the corroding metal and an ICP the polymer could be reduced causing dopants to be set free. If dopants with corrosion inhibiting properties are chosen, the release from the ICP might result in the inhibition of the corrosion in the defect and the delamination of the coating. Kendig et al. showed that, in the case of polyaniline, anions can be released by a corrosion induced increase in pH, resulting from oxygen reduction at the conducting polymer [12]. Nevertheless, no results showing successful dopant release as a consequence of electrochemical ICP reduction induced by the anodic corrosion reaction in the defect have been reported.

Recently, we could show that such an electrochemically triggered release is possible by adding polypyrrole nanoparticles to the organic coating [13]. The electrochemical reduction of the polypyrrole nanoparticles is caused by the decrease in potential that is a characteristic feature of electrochemically driven delamination. Only if corrosion occurs at a defect site this decrease in potential will take place. Hence, no unwanted and unnecessary leaching can occur. The release stops automatically when the corrosion is successfully stopped.

However, several criteria have to be taken into account for the preparation of a coating capable of this efficient intelligent release. These will be discussed in the following.

Experimental

Materials and sample preparation

Stable water dispersions of polypyrrole (Ppy), doped with either $S_2O_8^{2-}$, MoO_4^{2-} or $[PMo_{12}O_{40}]^{3-}$, a large polymolybdophosphate (with the phosphate at the centre and surrounded by the molybdate units, for detailed structure see [14]), have been prepared by chemical polymerization in the Institute of Macromolecular Chemistry and Textile Chemistry at the Technical University Dresden (Prof. H.-J. Adler) by Dr. A.Z. Pich and Dr. Y. Lu. The method has been described in detail elsewhere [15]. A Ppy dispersion doped with the $S_2O_8^{2-}$, which is not a corrosion inhibitor, has been used as a reference sample. For all experiments, iron sam×10 mm in size, were abraded up to 1200 grid paper, cleaned in acetone and then in ethanol in an ultrasound bath for 3 minutes. The Ppy dispersions have been mixed with a non-conducting matrix polymer provided by Chemetall GmbH Frankfurt/Main (unpigmented basis polymer OCD 860) and applied with the use of a spin-coater on iron samples. Subsequently, samples were cured at 100°C for 5 minutes. Additionally a clear top-coat polymer (BASF) has been applied.

Ppy films characterization

For well defined application the composite films were prepared by spin-coating. SEM (LEO 1550 VEP) investigations revealed that after three consecutive applications a compact, closed monolayer film was obtained on the iron surface. Further spin-coating of Ppy dispersion showed no changes in the SEM images. The investigations indicated that the Ppy coating is comprised of one Ppy core particle monolayer layer, distributed evenly on the iron surface. Hence, the thickness of the film is close to the diameter of Ppy cores i.e. from 70 to 120 nm. This conclusion is supported also by results of XPS sputtering and ellipsometry measurements.

Scanning Kelvin Probe (SKP) experiments

SKP delamination experiments

In order to perform the SKP experiment, part of the iron sample has been covered with the dispersion coating and on the other part an artificial defect has been prepared (Figure 1). In the defect 0,1 M KCl solution has been injected. After the SKP experiments, the delaminated Ppy coatings could be easily peeled off from the metal surface. XPS analysis of the delaminated Ppy films and the surface of the corroded iron samples was performed.

SKP reduction experiments

To evaluate the pH effect on the anion release, reduction experiments of Ppy coatings in a controlled N_2 atmosphere have been performed.

Electrochemical characterization

Cyclic voltammetry and reduction (polarization) experiments were carried out in a three-electrode cell in 50 ml of 0,1 M $NaClO_4$ or 0,1 M $C_{16}H_{36}NCl$, purged with N_2 for 30 minutes prior to the experiments, with a scan rate of 50 mV/s, using a HEKA potentiostat.

Results and Discussion

The electrochemical characterization of the composite monolayer coatings revealed that these composite coatings are electrochemically active, i.e. the non-conducting matrix polymer does not totally inhibit the electrochemical activity of the polypyrrole. For both coatings, $PpyPMo_{12}O_{40}$ (**PpyPMo**) and $PpyMoO_4$ (**PpyMo**), the reduction and oxidation peaks in aqueous 0.1 M $LiClO_4$ solution are similar. In 0.1M tetrabutylammonium chloride solution, where the large size of the cation makes its incorporation during reduction of the Ppy rather improbable, for PpyPMo the redox peaks change significantly their position after a few first cycles. That indicates that $[PMo_{12}O_{40}]^{3-}$ are systematically replaced with Cl^- anions, which are available in the solution in high concentration. For PpyMo no changes in the cyclovoltammograms are observed, as the plain molybdate and the chloride anion do not differ so dramatically in structure as do $[PMo_{12}O_{40}]^{3-}$ and Cl^-, but the observed reversibility in the CVs indicates that molybdate anions are released and chloride is incorporated. This is also verified by XPS analysis of the coatings after the electrochemical experiments.

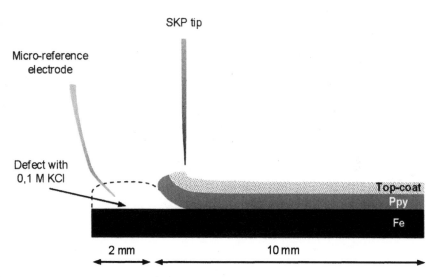

Figure 1. Sample set up for SKP experiments: Sample is an iron sheet, coated by the PPy-containing composite and a top-coat. The tip is scanned over the surface of the coating, and after calibration over a Cu/CuSO₄ reference it directly delivers the electrode potential at the polymer/metal interface. Starting from the defect a delamination front will proceed into the intact interface. At the delaminated interface the potential is pulled down to the defect potential, which is directly measure dby SKP (see e.g. [2]).

Hence, for both coatings the incorporated PPy is electrochemically active and the anions can be released by electrochemical reduction, i.e. their release during corrosive attack should in principle be possible.

The most dangerous consequence of corrosion in a defect site in the coating is a fast electrochemically driven delamination of the coating, driven by the corrosion in the defect. In the especially dangerous cathodic delamination the degradation at the polymer/metal interface is caused by radicals produced by oxygen reduction at the interface. While corrosion is efficiently inhibited at the interface, oxygen reduction can principally take place there. Ions, water and oxygen can diffuse in sufficient quantities through any organic coating. However, the isolated intact interface, the anodic counter reaction is missing, i.e. no electrons are available for the oxygen reduction. Hence, the potential at the intact interface stabilizes at quite high potentials, where oxygen reduction is kinetically inhibited. In the presence of a defect, however, the defect and the delamination site form anode and cathode of a galvanic element, which migrates cations along the delaminated interface. Thus a closed electric circuit for an electron flow from the defect to the polymer/metal interface is established (see figure 2 and e.g. [2]). The potential is pulled down and oxygen reduction takes place and further degrades the interface, facilitating cation transport and decreasing the i·R drop along the delaminated interface.

This potential decrease at the interface should reduce the polypyrrole in the prepared composite coatings and cause the release of the corrosion inhibiting anions.

Figures 3-4 show the delamination behaviour of the PpyPMo and PpyMo containing composite coatings. As can be seen at the upper right side of each diagram, the potential of the intact Ppy coatings remains between 0,1 and 0,2V vs. SHE. Profiles moving towards the right side of the diagram represent the propagation of the delamination front from the defect inside the intact polymer coating (see also figure 1). The border between defect and coating is at x=0. Intervals between the consecutive profiles are for both cases the same and equal 20 minutes. The composite coating containing Ppy doped with the MoO_4^{2-} anion delaminates fast, about a hundred micrometers in ten minutes (figure 3). No passivation of the defect could be observed and the delamination velocity remains constantly high. This is surprising as it was expected that molybdate anions should be released. Since molybdate is an extremely potent corrosion inhibitor on iron, even small amounts should show a significant effect. Other composite coatings containing Ppy doped with anions that have no corrosion inhibition effect, such as sulphate, show basically the same behaviour.

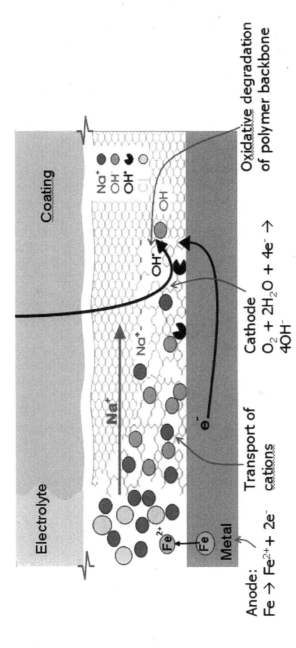

Figure 2. Sketch of the delamination process. Electrons can flow from the defect to the polymer/metal interface, if the electric circuit is closed by cation transport from the defect into this interface. Oxygen reduction takes place and the radicals produced as side products degrade the interface.

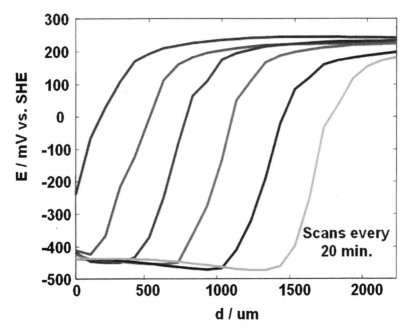

Figure 3. SKP delamination profiles (successive scans from left to right) for PpyMo. The defect border is at x=0. Left of this is the actively corroding defect, right of this the coated surface. The delamination speed remains constant throughout the measurement. Although molybdate is a powerful corrosion inhibitor no passivation of the defect can be observed, since the potential of the delaminated area near the defect stays at about -450 mV, the potential of actively corroding iron. For a composite coating with polypyrrole doped with sulphate, an anion which has no corrosion inhibiting properties, practically the same behaviour is observed. [13]

But in the case of Ppy doped with $[PMo_{12}O_{40}]^{3-}$ (figure 4) an untypical and interesting behaviour can be observed. The first profile does not differ from the ones measured in the case of the PpyMo sample but already the second one shows a significant potential increase at the delaminated area $\Delta E = 0,17V$. Consequently the potential difference between the delaminated area and the intact polymer coating, which is the driving force for further delamination, is getting smaller. Hence, we can observe that the delamination speed significantly decreases. The potential in the already delaminated area changes further and becomes even more positive with time. The change is in the case of the presented sample 0,28V and usually between 0,15V and 0,3V.

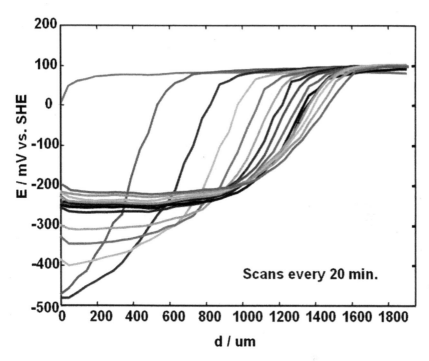

Figure 4. SKP delamination profiles for PpyPMo. Here, after about one hour of corrosive attack the corrosion potential in the defect (see potential near x=0) starts to shift to more positive potentials, indicating passivation of the defect. After a while a shift of nearly 300 mV is observed, which is substantial. Since the potential difference between the defect and the intact area is a driving force for the delamination, this decrease results in a significant slow down in delamination velocity. This slow down is the reason why full passivation of the defect is not achieved, because it causes a decrease of the amount of further released inhibitor anions. But this exactly is how this intelligent coating works: for its protection, no further passivation is required. Hence it stops further release. The final value of the potential at the delaminated area near the defect might indicate that oxygen reduction at the interface, the prerequisite for interface degradation, does not occur any at this potential. [13]

This observation is supported by the simultaneous measurement of the open circuit potential at the defect, using a micro-reference electrode [16], where the same significant anodic potential shift equal 0,17V has also been observed.

In the case of the PpyPMo containing composite coating, subsequent XPS analysis performed at the surface of the iron sample shows the presence of molybdate at the defect as well as, after the coating has been peeled off, at the delaminated area. The amount of molybdate is variable and changes in the range between 1 and 3% (atomic per cent) for individual experiments, a sufficient amount for a passivation effect. Small amounts of phosphate, usually ~1% have been also found at the metal surface. In the delaminated Ppy film K^+ can be found indicating that, parallel to the anion release reaction, to some extend, cation incorporation takes place. It has been already reported in literature [17,18] that during the reduction process of Ppy, not only the anion release can take place but for the charge compensation usually also cation incorporation is possible.

An equivalent XPS analysis for the case of the composite coating containing PpyMo, i.e. where the polypyrrole was doped with the plain molybdate, showed that in this case no anion release took place; only cation incorporation took place. This explains why the delamination behaviour for this coating is basically the same as for a coating containing polypyrrole doped with anions that cannot inhibit corrosion. This is surprising, as the plain molybdate is a quite small anion and its release should be much easier than the one of the larger polyphosphomolybdate. So how can this observation being explained?

One important difference in the two cases is that in the case of the large phosphopolymolybdate anion, a decomposition reaction should follow its release [19], as it is instable at higher pH typically prevailing at the delaminating interface (see e.g. [2]):

$$[PMo_{12}O_{40}]^{3-} \underset{H^+}{\overset{OH^-}{\rightleftarrows}} [PMo_{11}O_{39}]^{7-} + MoO_4^{2-} \underset{H^+}{\overset{OH^-}{\rightleftarrows}} [P_2Mo_5O_{23}H]^{5-} + MoO_4^{2-} \underset{H^+}{\overset{OH^-}{\rightleftarrows}} \left.\begin{matrix} HPO_4^{2-} \\ MoO_4^{2-} \end{matrix}\right\}$$

Obviously, the decomposition of the large polymolybdate buffers the pH at the interface.

And the pH may indeed play an important role in the processes observed here. According to some authors [20-22] OH$^-$ ions as strong nucleophiles can react with Ppy causing its overoxidation. Depending on the exact structure of the polypyrrole, this might cause anion release or anion trapping. Knowing that during the delamination experiments Ppy is exposed to OH$^-$ attack, in the case of PpyPMo less than in the case of PpyMo, consequently we might observe the anion release due to the OH$^-$ nucleophilic attack on Ppy and not a potential driven release. Hence, PpyPMo reduction experiments in controlled N_2 atmosphere, when OH⁻ influence on the release is excluded, have been performed. Also in this case XPS analysis showed the presence of molybdate at the iron surface. This result shows that $[PMo_{12}O_{40}]^{3-}$ anions are indeed released due to influence of potential decrease, i.e. as a consequence of true electrochemical reaction.

While the $[PMo_{12}O_{40}]^{3-}$ anion decomposition reaction provides a buffering effect, in the case of a PpyMo coating, the polyppyrole is exposed to much higher pH at the delaminating interface. That may cause a fast polypyrrole overoxidation process. It is possible that, depending on conditions and identity of Ppy dopants used, as a consequence of the overoxidation process, dopants will stay trapped in the non-conducting, overoxidised Ppy matrix [23].

To determine whether in the case of Ppy doped with plain MoO_4^{2-} anions, this undesired OH⁻ influence causes the observed anion immobilization in the coating, reduction experiments of the PpyMo coating in controlled N_2 atmosphere have been performed. The results of the performed XPS analysis indicate that during this experiment now the MoO_4^{2-} anion is indeed released from the polymer coating [13].

Experiments on pure, electropolymerized polypyrrole coatings doped with molybdate did not show anion release, either. In the case of pure polypyrrole coatings cation incorporation is generally much faster than anion release for the long distances involved in the delamination set up, [10; 24]. In fact, in general extended percolation networks of the conducting polymer have to be avoided in order to prevent this fast cation incorporation. This limits the amount of conducting polymer that can be added to the coating. On the other hand, a high amount is desirable as this ensures storage of a high amount of inhibitors.

Hence, the determination of the optimal distribution of the conducting polymer in the coating and at the interface is topic of current research.

Conclusions and Summary

Composite coatings containing polypyrrole particles doped by polyphosphomolybdate anions show a true electrochemically driven on demand release of inhibitor anions. The trigger for the release is the potential decrease that occurs during delamination at the polymer/metal interface. As this decrease will only and solely happen in presence of a corroding defect, the inhibitor anions are safely stored in the coating until the moment of corrosive attack. They then are released, until the corrosive attack is stopped or contained to non-dangerous levels. Then release is stopped or reduced to very low limits. Hence, this system can really be called a truly intelligent self-healing coating.

As the defects prepared for the experiments presented here are unusually large, this result is really remarkable. For more realistic smaller defects the effect should be excellent.

However, some criteria have to be taken into account. First of all, an increase of pH beyond certain levels may render the anions immobilized in an over-oxidized polypyrrole matrix. This problem can be solved by providing a buffer, e.g. by using polyphosphomolybdate anions instead of plain molybdate. Secondly, only composite coatings show this effect. Pure conducting polymer films fail in the presence of large defects, although they might be able to heal small pinholes. In fact, the exact amount and distribution of the conducting polymer plays a crucial role and is topic of current research.

Acknowledgements

The results reported here are part of the PhD work of G. Paliwoda. Special acknowledgements go to K. Potje-Kamloth, A.Z. Pich and Y. Lu, H.-J. Adler from the TU Dresden, Germany, for providing the polypyrrole core particles doped with different anions.

References

1. B. Baumert, M. Stratmann, M. Rohwerder, *Z. f. Metallkunde* **2004**, *95*(6), 447-455.
2. M. Rohwerder, M. Stratmann, *MRS Bulletin* **1999**, *24* , 43-47.
3. D.W. De Berry, *J. Electrochem. Soc.* **1985**, *132*, 1022.
4. G.M. Spinks, A.J. Dominis, *J. Solid State Electrochem.* **2002**, *6*, 85-100.
5. Wessling B., *Materials and Corrosion* **1996**, *47*, 439-445.
6. J. Reut, A.Öpik, K. Idla, *Synthetic Metals* **1999**, *102*, 1392-1393.
7. T. D. Nguyen, M. Keddam, H. Takenouti, *Electrochem.and Solid-State Letters* **2003**, *6*, B25-B28.
8. P.J. Kinlen, D.C. Silverman, C.R. Jeffereys, *Synthetic Metals* **1997**, *85*, 1327-1332.
9. T. Schauer, A.Joos, L.Dulog, C.D. Eisenbach, *Progress in Organic Coatings* **1998**, *33* 20-27.
10. A. Michalik, M. Rohwerder, *Z. f. Physik. Chemie* **2005**, *219*(11), 1547-1560.
11. J.N. Barisci, T.W. Lewis, G.M. Spinks, C.O. Too, G.G. Wallace, *J. Intel. Mat. Syst. Str.* **1998**, *9*, 723.
12. M. Kendig, M. Hon, L. Warren, *Prog Org Coat.* **2003**, *47*, 183.

13. G. Paliwoda, M. Stratmann, M. Rohwerder, *et al.*, *Corros. Sci.* **2005**, *47* 3216-3233.
14. J. C. A. Boeyens, G. J. McDougal and J. van R. Smit, *J. Solid State Chem.* 18 (1976) 191-199.
15. Y. Lu, *Polypyrrole-containing composite particles: preparation, characterization and application.* 2005, TU/ Dresden: Dresden.
16. A.W. Hassel, K. Fushimi, M. Seo, *Electrochem Commun*, **1999**, *1*, 180.
17. J.Y. Lim, W.K. Paik, I.H. Yeo, *Synthetic Met*, **1995**, *69*, 451.
18. S.J. Dong, G.H. Lian, *J Electroanal Chem*, **1990**, *291*, 23.
19. G.A. Tsigdinos, *Top. Curr. Chem*, **1978**, *76*, 1.
20. Y.F. Li, R.Y. Qian, *Electrochim Acta*, **2000**, *45*, 1727.
21. H.L. Ge, G.J. Qi, E.T. Kang, K.G. Neoh, *Polymer*, **1994**, *35*, 504.
22. F. Beck, P. Braun, M. Oberst, *Ber Bunsen Phys Chem*, **1987**, *91*, 967.
23. F. Palmisano, C. Malitesta, D. Centonze, P.G. Zambonin, *Anal Chem.* **1995**, *67*, 2207.
24. A. Michalik and M. Rohwerder, submitted to *Electrochimica Acta*.

Chapter 15

Investigations of an Electrochemical Basis for the Protection of Steel and Aluminum by Polyaniline and Polyphenylene Ether Coatings

Peter Spellane

Department of Chemistry, New York City College of Technology–CUNY,
300 Jay Street, Brooklyn, NY 11201–1909

The use of inherently conductive polymers (ICPs) in anti-corrosion coatings for metals is presented with emphasis on polyaniline (PANI); a mechanism for metal protection, based on the oxidation – reduction chemistry of PANI and the passivity of steel or aluminum, is described. We have also examined the non-conductive polymer poly(2,6-dimethyl-phenylene ether) (PPE) and present evidence that PPE applied as thin coatings on aluminum coupons (Al 2024T3 and Al6061T6) enhances the metal's resistance to corrosion. Enhanced corrosion resistance is evident in salt-fog exposure test data and supported by DC anodic polarization curves of PPE-coated aluminum. We propose that coatings that comprise either PANI or PPE alter the surface chemistry of active-passive metals in a way that enables the metals to form better, more protective oxides, that is, that PANI and PPE protect substrate metals through a form of anodic protection. Recent reports concerning the use of ICPs in metal protection are surveyed, and new ideas for ICP-based smart coatings are presented.

Inherently Conductive or "Conjugated" Polymers in Coatings

The 2000 Nobel Prize for Chemistry was awarded to Professors Alan Heeger, Alan MacDiarmid, and Hideki Shirakawa for "the discovery and development of conductive polymers," materials that have enormous commercial potential (*1-4*). In retrospect, Heeger, MacDiarmid, and Shirakawa's process of invention seems simple: combining their expertises, they prepared *trans*-polyacetylene, the simplest macromolecule that has alternating single and double "conjugated" π-bonds, and added halogen to oxidize the polymer. Removing one or several electrons from the delocalized π-orbitals created mobile charge carriers. Polymers of this kind are called "inherently" conductive because their carriers of electronic charge move in molecular orbitals of the polymer itself rather than in conduction bands of additives such as graphite or metal flake.

This paper concerns the use of inherently conductive polymers (ICPs) in a large and environmentally-important application, as active ingredient in protective coatings for metals, that is, as potential replacements for the widely used but hazardous hexavalent chromium corrosion inhibitors. We describe how ICPs can react with substrate metal and ambient oxygen to function in smart protective coatings for metals. We review the history of polyaniline coatings on metals, report our measurements of electrochemical effects of polyaniline coatings on steel and both electrochemical and salt fog testing of polyphenylene ether coatings on aluminum. We review work from several other labs toward development of smart ICP coatings for metals. The ICPs have been well-documented (*5-6*) and their significance as smart materials recently surveyed (*7*). A review of corrosion control by ICPs, including accounts of important initial investigations, appeared in 2003 (*8*).

Commercially-Produced ICPs and Metal Protection

Polyacetylene (PA) is the simplest of the ICPs, the easiest to draw but the most difficult to handle. PA is unstable in air and is not in any commercial sense important. Conjugated resins that are easily-handled, produced on commercial scale, and finding use in various high-value applications include polyaniline (PANI), substituted-polythiophenes (PT), and polyphenylene vinylenes (PPV). Molecular structures of fragments of these are represented in Figure 1. On its rediscovery as a conductive polymer, physicists and chemists examined PANI. PANI is easy to prepare, but it is a complex material, with several accessible oxidation states and acid–base chemistry. "Emeraldine" polyaniline has approximately equal numbers of imine and amine nitrogen atoms; polyaniline-emeraldine base (PANI-EB) is represented in Figure 1.

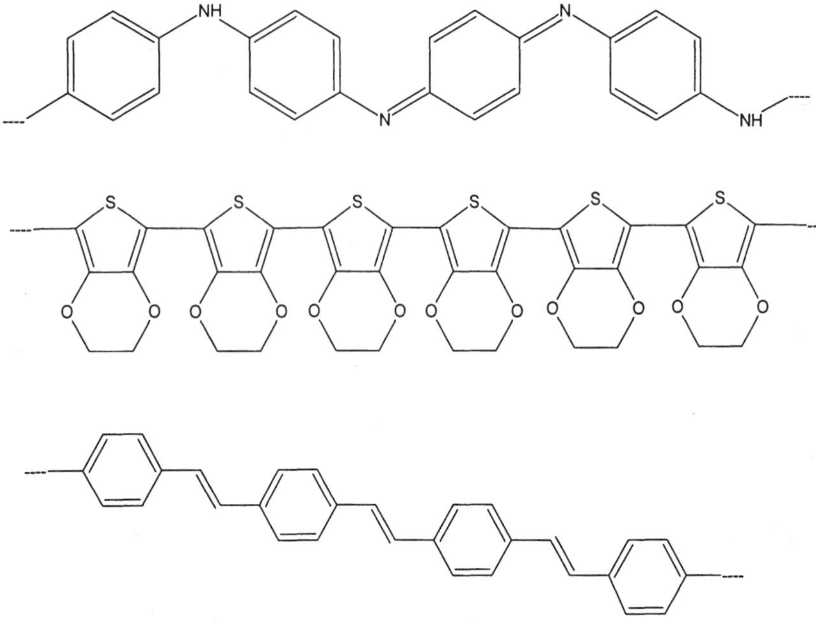

Figure 1. Fragments of commercially available ICPs, polyaniline (upper), polythiophene (middle), and polyphenylene vinylene (lower).

Addition of H_2 to each diimino-quinone group in the polymer yields the fully reduced "leucoemeraldine" polyaniline. The electrically-conductive form of PANI is prepared by protonation of PANI-EB. Applications for conductive polyaniline include antistatic coatings and conductive inks and adhesives. Several polythiophenes, with different pendant groups, are available. The poly(3,4-ethylenedioxythiophene) (PEDOT), shown here in its non-oxidized form, is produced by H. C. Starck, a division of Bayer Material-Science. Applications for PT include use in antistatic coatings and high-conductivity coatings, and as components of organic light emitting diodes, luminescent materials, and organic field effect transistors. Polyphenylene vinylene (PPV) can form highly ordered crystalline films; its applications include as electroluminescent material in organic light emitting diodes.

The instability of metals in air, their tendency to react with ambient oxygen to form metal oxides, creates a problem to producers and users of metal parts and an opportunity for chemists and electrochemists. Essentially every metal part intended for structural or functional use is coated for protection, and "coating" is almost invariably a multi-step process. Metal parts are usually pretreated (prepared for painting) with protective chemical conversion coatings, which are formed *in situ* in the chemical reactions of acid salts with metal surfaces. Phosphate conversion coatings are formed when for example steel is sprayed with a dilute zinc or manganese acid orthophosphate solution. A conversion coating may be saturated with oils or corrosion inhibitors or used as formed to promote the adhesion of paint. Zinc or zinc-aluminum surfaces are often treated with chromate coatings, formed when the metal is immersed in an acidified chromate solution. The paints that coat most metal articles can provide protection in several ways: barrier coatings prevent oxidants from contacting substrate metal; cathodic paint formulations provide zinc or aluminum metal as sacrificial anode to maintain the integrity of substrate metal; inhibitive paint coatings, typically comprising $SrCrO_4$ or other inorganic oxidant, react with substrate metal to maintain a protective metal oxide surface.

"Active-passive" metals, described more fully below, can form metal oxide surfaces that protect bulk metal. These include some metals of the greatest commercial importance: steel, copper, titanium, and aluminum. In principle, electrochemically-active polymer coatings can be engineered to oxidize active-passive metals and promote the reactions that lead to formation of protective metal oxide surfaces.

ICPs may also be useful as sensors components of smart coatings: because corrosion and metal protection involve redox chemistry, ICPs could be designed to monitor the integrity of substrate metal, generating observable signals from electrochemical changes taking place at the surface of the metal.

Polyaniline: Indications of Metal Protecting Effects

Polyaniline is among the earliest of industrial polymers, a by-product of the aniline dye industry in the 19th century, and probably the very first electroactive polymer. In 1862, a London physician reported electrochemical deposition of "a thick layer of dirty bluish-green pigment" on a platinum anode immersed in an acidic solution of aniline (9). The pigment was not identified as polyaniline, which was yet to be described, but the details in the report make almost certain that it was indeed PANI. The material's preparation, acid-base behavior, and redox properties were noted. Thus, one finds both a sophisticated *in situ* preparation of a new coating material and a description of the material's "smart" chemistry in a report that predates by about 40 years an understanding of the material's molecular nature (*10-13*) and by more than a century an examination of its efficacy in metal protection.

D. W. DeBerry reported in 1985 that stainless steels might be protected from acid corrosion by polyaniline coatings (*14*). DeBerry proposed, "A form of anodic protection may be obtained by coating an active/passive metal with a redox species capable of maintaining the native oxide on the metal." DeBerry had electro-coated stainless steel electrodes in perchloric acid with polyaniline and identified the "smart" character of the material: the electroactive form of

PANI appeared to be continuously regenerated by oxygen, a redox chemistry that could enable the long-term stability of PANI-coated metal.

Examination of polyaniline in anti-corrosion coatings intensified after 1991. A NASA Conference Paper (*15*) was followed by reports from W-K Lu, R. L. Elsenbaumer, and B. Wessling (*16-17*), Yen Wei at Drexel University (*18*), Arthur Epstein and colleagues at the Ohio State University (*19-20*), and from our laboratory at Akzo Nobel Chemicals (*21*). Evidence of oxidation-reduction chemistry between polyaniline and substrate metals supported DeBerry's earlier report. XPS data indicate that emeraldine PANI, in either base or protonated form, can be reduced by substrate steel to the leucoemeraldine state and re-oxidized by air to the emeraldine state, suggesting that a polyaniline coating could enhance an active-passive metal's ability to react with oxygen to form a metal-protecting oxide surface, as indicated in Figure 2.

A Monsanto group proposed an alternative mechanism for metal protection by ICPs (*22*): ICPs shift the site of oxygen reduction from the metal-coating interface into the bulk polymer, thereby reducing the delamination of coatings caused by reduction products.

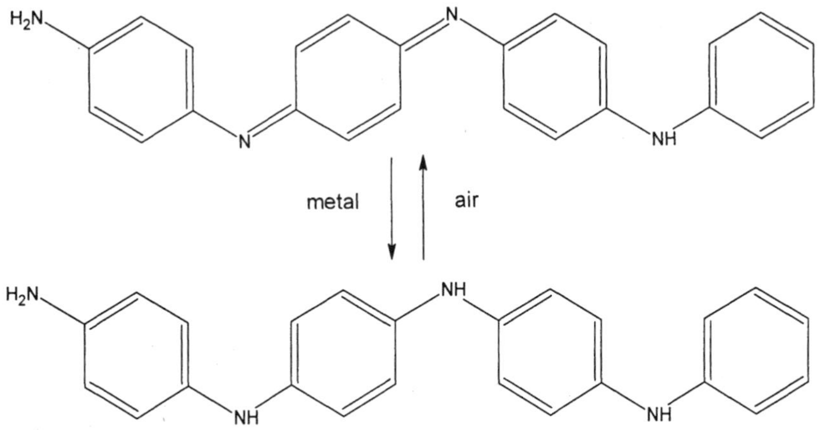

Figure 2. Reduction by metal of polyaniline-emeraldine base (PANI-EB) to polyaniline-leucoemeraldine base (PANI-LEB), indicated by arrow down, and air oxidation of PANI-LEB to PANI-EB, indicated by up arrow.

Recent Work Concerning ICPs in Metal-Protecting Coatings

Much of the newest work in the development of ICP-based anti-corrosion coatings for metals involves hybrid systems, coating formulations that combine a functional ICP with another agent of metal protection. The dopants that enable electrical conductivity in ICPs affect the polymers' anti-corrosion performance, an effect that has been described and reviewed by Wallace and colleagues (23). Kinlen and colleagues reported (24) that phosphonic acid-dopant enhances polyaniline's anti-corrosion efficacy. Both Kendig (25) and Kinlen (26) have reported the controlled release of organic inhibitor, triggered by changes in potential at surfaces of polyaniline-coated copper and copper-rich aluminum, provides metal protection. Rohwerder and coworkers reported electro-chemically-triggered release of inhibitor achieved by incorporation of polypyrrole nanoparticles in coating formulations (27).

Several recent patents claim that ICPs enhance the cathodic protection afforded by coatings that incorporate sacrificial anodes such as zinc or aluminum (28-29). Like some of the earliest reports of polyaniline on metal, with PANI coatings formed electrochemically, some very recent reports also describe metal protection with ICP coatings formed electrochemically on metal surfaces (30-34).

Metal Passivation: Spontaneous Metal Protection by Metal Oxide Coatings

In the process of passivation, "active-passive" metals spontaneously form surface oxides that, to some degree, adhere to bulk metal and protect the metal from further oxidation. Descriptions of the process can be found in the corrosion engineering textbooks (35-37). Metal passivation is the phenomenon that slows the thermodynamically-favored oxidations of steel, aluminum, and titanium and, in some cases, enables even uncoated metals to maintain structural integrity for years. Metal passivation can be driven electrochemically as for example in the "anodization" of aluminum. The passivation of an active-passive metal is demonstrated in DC potentiodynamic current-voltage curves. As increasingly positive (anodic) voltages are applied to an active-passive substrate metal, current between the working and counter electrodes varies over several orders of magnitude, producing a current-voltage curve that approximates the idealized I-V curve shown in Figure 3. As bias voltage is scanned anodically, current density finds a minimum at the open-circuit potential (E_{OC}). Current density increases exponentially at bias positive of E_{OC}, until it changes abruptly at the "passivating potential" (E_{PP}), indicating a transition from the metal's active to passive regimes. In the passive region, the metal resists further oxidation, even under increased bias, as indicated by the nearly constant current density. At very strong anodic bias, the oxide degrades, and the metal undergoes rapid oxidation in its "transpassive" region.

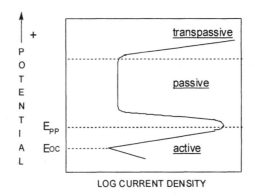

Figure 3. Idealized current voltage curve indicating active-passive behavior of some metals. E_{OC} and E_{PP} are open circuit and passivating potentials. Active, passive, and transpassive regions are indicated.

Experimental Section: DC Potentiodynamic Measurements of Polymer-Coated Steel or Aluminum and Salt Fog Corrosion Testing of PPE-Coated Aluminum

Preparation of PANI or PPE Coatings on Metal Samples

Steel and aluminum coupons were coated with polyaniline-emeraldine base (PANI-EB) or with poly(2,6-dimethylphenylene ether) (PPE) for evaluation of each resin's effectiveness in protecting metals. PANI-EB was prepared in lab in the ammonium persulfate oxidation of aniline. In the potentiodynamic tests reported here, PANI-EB powder had been added as a pigment to a coating formulation, dispersed mechanically in a conventional epoxy-melamine primer.

Cold rolled steel coupons were obtained from ACT Laboratories, Inc. Hillsdale, MI 49242. Metal coupons were scrubbed with 400 grit sandpaper and wiped with acetone or 2-butanone-saturated paper towel.

Aluminum coupons (Al6061T6, 03x06x0.032 inches, cut only unpolished, product number APR20754, and Al2024T3 03x06x0.025 inches, cut only unpolished, product number APR24796) were obtained from ACT Laboratories, Inc. The coupons were cleaned with acetone, abraded with 400 grit sandpaper, wiped again with acetone, then bar-coated with a solution of PPE, 10% (w/w) in toluene.

The poly(2,6-dimethylphenylene ether) solutions were prepared by adding 10 grams GE Blendex BHPP820 (GE Specialty Chemicals, now Chemtura, Parkersburg, WV) slowly to 90 grams toluene. With heating, gold colored solutions were obtained. The PPE solutions were filtered through 0.45 µ PTFE or polyethylene filters and barcoated on the coupons with a #24 wire-wound rod. The coated samples were dried at in air at room temperature or baked in ovens for 10 minutes at 90-95 °C or 200-210 °C then cooled by immersion in water. Coatings prepared in this way were determined, with an Elcometer 300 coating thickness gauge, to have thicknesses between 1 and 3 μm.

Potentiodynamic Current—Voltage Curves

In earlier work, we adapted DC potentiodynamic methods, typically used to characterize uncoated metal alloys, to characterize metal surfaces coated with formulations that did or did not contain chromate inhibitors (38). We have used the same technique to study metal coated with ICPs as neat coatings or in formulations. In these measurements, coated surfaces were scored with a knife-edge to ensure exposure of approximately 1 cm^2 of metal surface to electrolyte solution. The experimental configuration is diagrammed in Figure 4.

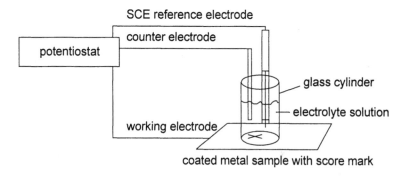

Figure 4. Diagram of electrochemical cell used to measure DC potentiodynamic current – voltage curves for coated metal substrates.

Salt Fog Corrosion Testing

To examine the performance of polyphenylene ether resin as an anti-corrosion coating, we have at various times coated steel, copper, titanium, and aerospace aluminum by spray- or bar-coating substrate metal with toluene solutions of the resin. In other studies, we added PPE as a pigment to standard primer formulations.

Salt fog testing (ASTM B 117) and evaluations of aluminum panels coated with thin neat PPE, were performed at DL Laboratories (New York, NY). Panels were observed daily for 5 days and evaluated for blistering (ASTM D 714), corrosion (visual inspection), and salt deposition (visual inspection), and graded according to this ASTM score system:

Rating	Performance	Effect
10	perfect	none
9	excellent	trace
8	very good	very slight
6	good	slight
4	fair	moderate
2	poor	considerable
0	no value	complete failure

NOTES:
*   Cannot be ascertained because of severity of corrosion and/or salt deposit.
**   Blister area observed as a corrosion site.
9F   Type 9 blister, frequency few.

Results Section: Electrochemical and Performance Testing of PANI-EB or PPE as Coatings on Steel or Aluminum

Potentiodynamic Measurements of PANI-EB-coated Cold Rolled Steel

Using a DC potentiodynamic method to characterize the passivity of coated steel (38), we measured the current-voltage curves for cold rolled steel coated with either clear epoxy-melamine primer or epoxy-melamine primer with ca. 10% polyaniline-EB. The curves shown in Figure 5 suggest that polyaniline-EB fortifies the native passivity of cold rolled steel: the I-V curve of the PANI-coated steel sample has a more distinct passive region than does the steel that is coated with "clear" (unpigmented) epoxy-melamine. We infer from these measurements that polyaniline changes the metal's surface properties. While the polyaniline coating appears to fortify the metal's passivity, this effect alone does not indicate enhanced corrosion resistance, which depends on the values of open circuit potential and current density.

Salt Fog and DC Potentiodynamic Measurements of PPE-coated Aluminum

PANI-EB is a non-conductive form of polyaniline. Having observed that PANI-EB appeared to improve steel's passivity, we reasoned that the origin of polyaniline's action could be the polymer's redox chemistry rather than its electrical conductivity; this reasoning led us to test polyphenylene ether (PPE) as a protective coating. Poly(2,6-dimethylphenylene ether) is not conductive; the material finds commercial application mainly as a resin for thermoplastics (39). It blends well with polycarbonate, polystyrene, and other commercial resins; it is used in plastic liners and housings for electronics including computers

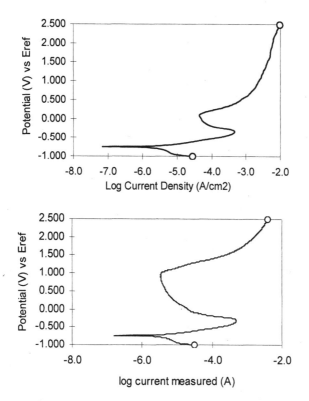

Figure 5. DC potentiodynamic I-V curves for steel coated with clear epoxy melamine primer (upper) or with epoxy melamine with PANI-EB (lower).

and televisions, kitchen and personal care appliances, and numerous automotive applications. Several early (*40-43*) and more recent (*44-47*) reports describe metals coated with PPE resins or claim PPE's effectiveness in protecting metal.

The molecular structures of polyaniline-leucoemeraldine (PANI-LEB) and poly(2,6-dimethylphenylene ether) are compared in Figure 6. Note that oxidation of either polymer can be reversed by reduction by the metal.

Results of ASTM B 117 testing of Al6061T6 and Al2024T3 aluminum panels coated with neat poly(2,6-dimethylphenylene ether) are presented here; blistering, corrosion, and salt deposition data appear in Tables 1, 2, and 3. Results presented in Table 2 have appeared in an earlier form (*47*).

The thicknesses of coating on panels whose tests are reported here had not been measured, but the coating of an identically-prepared Al2024T3/PPE sample was determined with an Elcometer 300 Coating Thickness Gauge, to be 2.30 +/- 1.0 micrometer.

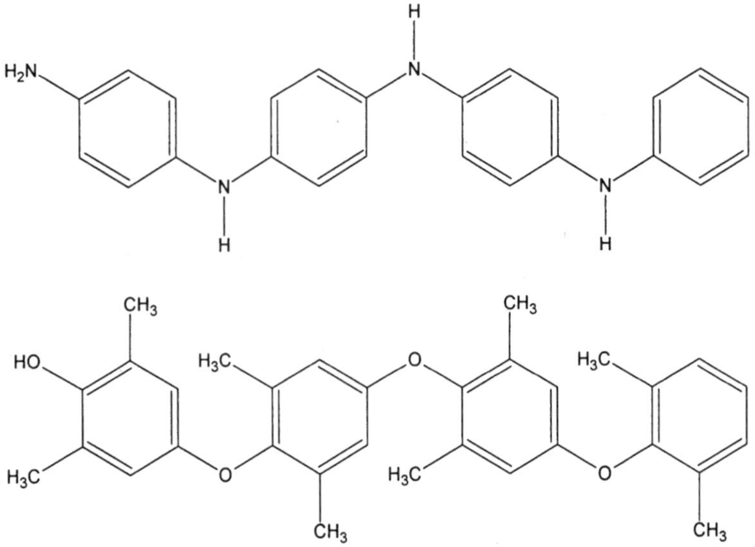

Figure 6. Comparison of the molecular structures of leucoemeraldine—polyaniline and poly(2,6-dimethylphenylene ether).

Table 1. BLISTERING During ASTM B117 Exposure

		Bake temp	Day 1	Day 2	Day 3	Day 4	Day 5
Al 6061T6	No coating	RT	10	*	*	*	*
	PPE	90C	9F	**	**	**	**
	PPE	90C	9F	**	**	**	**
	PPE	204C	10	10	10	10	10
	PPE	204C	10	10	10	10	10
Al 2024T3	No coating	RT	*	*	*	*	*
	PPE	90C	10	9F	9F	9F	9F
	PPE	90C	10	10	10	10	10
	PPE	204C	10	10	10	10	10
	PPE	204C	10	10	10	10	10

Table 2. CORROSION during ASTM B117 Exposure

		Bake temp	Day 1	Day 2	Day 3	Day 4	Day 5
Al 6061T6	No coating	RT	8	4	2	0	0
	PPE	90C	10	9	9	9	8
	PPE	90C	10	9	9	8	8
	PPE	204C	9	9	9	8	8
	PPE	204C	9	9	9	8	8
Al 2024T3	No coating	RT	0	0	0	0	0
	PPE	90C	9	9	8	8	6
	PPE	90C	9	9	9	8	6
	PPE	204C	9	9	8	8	6
	PPE	204C	8	8	8	8	6

SOURCE: Reproduced with permission from reference 47. Copyright 2000 Electrochemical Society, Inc.)

Table 3. SALT DEPOSITION during ASTM B117 Exposure

		Bake temp	Day 1	Day 2	Day 3	Day 4	Day 5
Al 6061T6	No coating	RT	6	2	0	0	0
	PPE	90C	10	10	10	10	10
	PPE	90C	10	10	10	10	10
	PPE	204C	10	10	10	10	10
	PPE	204C	10	10	10	10	10
Al 2024T3	No coating	RT	0	0	0	0	0
	PPE	90C	10	10	10	8	6
	PPE	90C	10	10	10	9	9
	PPE	204C	10	10	8	8	8
	PPE	204C	10	9	8	8	8

On the basis of the resin's structural similarity to polyaniline-LEB and the air-instability of each polymer, we propose that 1-electron oxidation of the terminal dimethylphenol to a quinone-like end-group allows the PPE polymer to act as an active and renewable agent of metal passivation. Note that a similar quinoid radical is proposed to be an intermediate in the oxidative polymerization that leads to formation of polymer. A mechanism of protection is proposed in Figure 7.

DC potentiodynamic measurements of aerospace aluminum Al2024T3 coated with thin, neat PPE coatings produced current-voltage curves that suggest that PPE enhances the metal's passive state: the open circuit potential is higher, the onset of the transpassive region at higher bias voltage, and current density in the transpassive region lower, in the PPE-coated sample than in the epoxy-coated sample. See Figure 8.

On the basis of these potentiodynamic current-voltage data, and with the support of salt fog performance data, we propose that coatings that comprise either PANI or PPE can alter the surface chemistry of active-passive metals in a way that enables the metals to form better, more protective surface oxides.

Work in Progress: Even Smarter ICP Coatings

Because ICP coatings successfully combine what have been traditionally distinct functions of metal coatings, adherent protective barrier and anti-corrosion agent, and because the polymers can be chemically modified and offer environmental advantage, it is reasonable to call them "smart." Indeed, in some of the newest formulations, coatings that contain ICPs use the polymer's response to environmental conditions to trigger more aggressive protection by other components: very smart.

A very smart and actively-protective coating would somehow remediate its imperfections or perceive and report any incipient problems. The latter of these smart functions is the goal of our current work: we seek to develop sensor compounds that can monitor coating performance. Our goal is optically- or electronically-addressable low-cost coatings that can report the early stages of metal oxidation. We have begun by dispersing porphyrin compounds in coatings.

The porphyrins are a family of stable macrocyclic aromatic compounds whose optical properties are well-characterized. Porphyrins and metal-chelated metalloporphyrins have highly characteristic optical absorption and emission properties. For example, the presence of a metal ion, such as Zn^{++}, can be verified by the optical properties of the metalloporphyrin compound it forms.

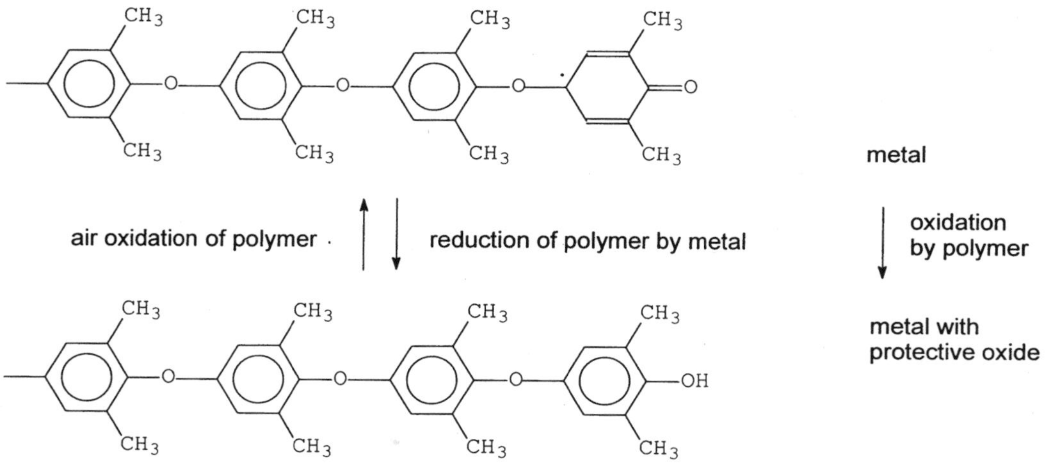

Figure 7. A proposed mechanism for metal protection: air oxidation of PPE coating on metal substrate creates a quinoid radical that can oxidize the metal.

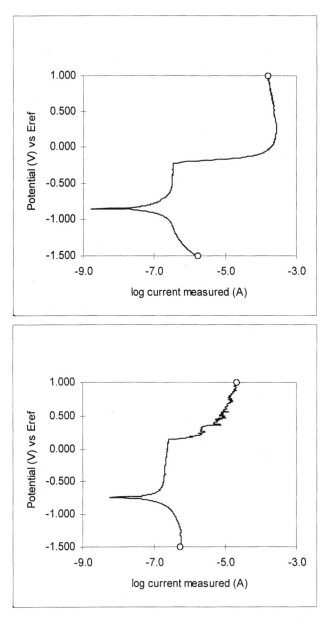

Figure 8. DC Potentiodynamic current – voltage curves, Al2024T3 coated with clear epoxy (upper) or with neat PPE (lower).

In recent years, porphyrin compounds have been used in smart coating applications. Luminescent metalloporphyrin compounds, particularly the tetra-perfluorotetraphenylporphyrin (TFPP) compounds, have been developed as optical sensors of airflow. The intensity and lifetime of optical luminescence of these porphyrins are inverse functions of O_2 pressure (48). These porphyrin-based "pressure paints" are smart coatings that are used in aerospace research and development to monitor airflow over aircraft surfaces.

Our current research involves synthetic chemistry: we want to bond conjugated sidechains to porphyrin compounds, to integrate optically-observable porphyrins into the electronic structures of ICP coatings. Porphyrin compounds with conjugated sidechains are also investigated for their ability to separate charge or transfer excited state energy under photo-excitation. These porphyrin-conjugated oligomer compounds hold promise of being materials with extraordinary stability and utility in electronic applications (49-51).

References

1. Shirakawa, H.; Louis, E. J.; MacDiarmid, A. G.; Chiang, C. K.; Heeger, A. G. *Chem. Commun.* **1977**, 578-580.
2. Shirakawa, H. *Angew. Chem. Int. Ed.* **2001**, *40*, 2574-2580.
3. MacDiarmid, A. G. *Angew. Chem. Int. Ed.* **2001**, *40*, 2581-2590.
4. Heeger, A. G. *Angew. Chem. Int. Ed.* **2001**, *40*, 2591-2611.
5. *Handbook of Conducting Polymers, 2nd Edition;* Skotheim, T. A.; Elsenbaumer, R. L.; Reynolds, J. R., Eds.; Marcel Dekker, Inc.: New York, 1998.
6. *Handbook of Organic Conductive Molecules and Polymers;* Nalwa, H. S., Ed.; Wiley: Chichester; New York, NY, 1997; Vol, 2, 3, 4.
7. *Conductive Electroactive Polymers: Intelligent Materials Systems;* Wallace, G. G., Ed.; CRC Press: Boca Raton, FL, 2003.
8. *Electroactive Polymers for Corrosion Control;* Zarras, P.; Stenger-Smith, J. D.; Wei, Y., Eds., ACS Symposium Series 843, American Chemical Society: Washington, DC 2003.
9. Letheby, H. *J. Chem. Soc.* **1862**, *15*, 161-163.
10. Goppelsroeder, F. *Internationale Elektrotechniche Ausstellung* **1891**, *18*, 978.
11. Goppelsroeder, F. *Internationale Elektrotechniche Ausstellung* **1891**, *19*, 1047.
12. Green, A. G.; Woodhead, A. E. *J. Chem. Soc.* **1900**, *97*, 2404.
13. Willstaetter, R.; Moore, C. W. *Ber.* **1907**, *40*, 2665.
14. DeBerry, D. W. *J. Electrochem. Soc.* **1985**, *132*, 1022.
15. Thompson, K. G.; Bryan, C. J.; Benicewicz, B. C.; Wrobleski, D. A. *Proceedings of Technology 2001, The Second National Technology Transfer Conference and Exposition,* NASA Conf. Pub. 3161; 1991; Vol. 1, pp. 339-347.

16. Wessling, B. *Adv. Mater.* **1994**, *6*, 226.
17. Lu, W-K; Elsenbaumer, R.; Wessling, B. *Synth. Met.* **1995**, *71*, 2163.
18. Wei, Yen; Wang, Jianguo; Jia, Xinru; Yeh, Jui-Ming; Spellane, P. *Polym. Mater. Sci. Eng.* **1995**, *72*, 563-564.
19. Jasty, S.; Epstein, A. J. *Polym. Mater. Sci. Eng.* **1995**, *72*, 565.
20. Fahlman, M.; Jasty, S.; Epstein, A. J. *Synth. Met.* **1997**, *85*, 1323.
21. Beard, B.; Spellane, P. *Chem. Mater.* **1997**, *9*, 1949-1953.
22. Kinlen, P. J.; Silverman, D. C.; Jeffereys, C. R. *Synth. Met.* **1997**, *85*, 1327.
23. Wallace, G. G.; Dominis, A.; Spinks, G. M.; Tallman, D. E. In *Electroactive Polymers for Corrosion Control;* Zarras, P.; Stenger-Smith, J. D.; Wei, Y., Eds., ACS Symposium Series 843, American Chemical Society: Washington, DC 2003; pp 103-123.
24. Kinlen, P. J.; Ding, Y.; Silverman, D. C. *Corrosion* **2002**, *58*, 490.
25. Kendig, M.; Hon, M. *Corrosion* **2004**, *60*, 1024.
26. Kinlen, P. J.; Graham, C. R.; Ding, Y. *Polymer Preprints* (ACS Division of Polymer Chemistry) **2004**, *45*, 146, and Abstracts of Papers, 228th ACS National Meeting, Philadelphia, PA, United States, August 22-26, 2004 (2004).
27. Paliwoda, G.; Stratmann, M.; Rohwerder, M. *Corrosion Sci.* **2005**, *47*, 3216. See also following paper in this volume.
28. Geer, S. K.; Hawkins, T. R. U. S. Patent 6,440,332, 2002
29. Geer, S. K.; Hawkins, T. R. U. S. Patent 6,627,117, 2003.
30. Oezyilmaz, A. T.; Kardas, G.; Erbil, M.; Yazici, B. *Appl. Surf. Sci.* **2005**, *242*, 97-106.
31. Oezyilmaz, A. T. *Surf. Coatings Tech.* **2006**, *200*, 3918-3925.
32. Iroh, J. O.; Gajela, P; Cain, R.; Nelson, T.; Hall, S. *International SAMPE Symposium and Exhibition,* **2005**, *50*, 117-130.
33. Breslin, C. B.; Fenelon, A. M.; Conroy, K. G. *Materials Design* **2005**, *26*, 233-237.
34. Huerta-Vilca, D.; Siefert, B.; Moraes, S.R.; Pantoja, M. F.; Motheo, A. J. *Mol. Crystals Liq. Crystals* **2004**, *415*, 229-238.
35. Jones, D. A. *Principles and Prevention of Corrosion, Second Edition,* Prentice Hall: Upper Saddle River, NJ, 1996.
36. Fontana, M. G. *Corrosion Engineering, Third Edition,* McGraw-Hill, Inc.: New York, NY, 1986.
37. Uhlig, H. H.; Revie, R. W. *Corrosion and Corrosion Control, An Introduction to Corrosion Science and Engineering, Third Edition,* John Wiley & Sons: New York, NY, 1985.
38. Spellane, P. *Prog. Org. Coatings* **1999**, *35*, 277-282.
39. White, D. M.; Cooper, G. D. In *Kirk-Othmer Encyclopedia of Chemical Technology, Third Edition,* Grayson, M., Ed.; John Wiley & Sons: New York, NY 1982; Vol. 18, pp. 594-615.
40. Schmukler, S. U. S. Patent 3,396,146, 1969.
41. Davis, H. R.; Taylor, C. W. U. S. Patent 3,455,736, 1969.

42. Pham, M-C.; Mourcel, P.; Lacaze, P-C.; DuBois, J-E. *Bull. Soc. Chim. France*, **1985**, 1169.
43. Otero, T. F.; Ponce, M. T. *Thin Solid Films* **1988**, *162*, 209.
44. Spellane, P.; Yahkind, A.; Abu-Shanab, O. U. S. Patent 6,004,628, 1999.
45. Spellane, P. U. S. Patent 6,376,021, 2002.
46. Guo, H.; Zarnoch, K. P. U. S. Pat. Appl. Publ. US 2002098366 A1 20020725, 2002.
47. Spellane, P. In *Corrosion and Corrosion Prevention of Low Density Metals;* Buchheit, R. G.; Shaw, B. A.; Moran, J. P., Eds.; Proceedings Volume 2000-23, Phoenix, Arizona, Fall 2000; The Electrochemical Society: Pennington, NJ, 2000; pp. 136-145. Reproduced by permission of the Electrochemical Society, Inc.
48. Kavandi, J.; Callis, J.; Gouterman, M.; Khalil, G.; Wright, D.; Green, E.; Burns, D.; McLachlan, B. *Review of Scientific Instruments* **1990**, *61*, 3340.
49. Seta, P.; Bienvenue, E.; Moore, A.; Moore, T.; Gust, D. *Nature* **1985**, *316*, 653-655.
50. Andreasson, J.; Kodis, G.; Ljungdahl, T.; Moore, A. L.; Moore, T. A.; Gust, D.; Mrtensson, J.; Albinsson, B. *J. Physical Chem. A* **2003**, *107*, 8825-8833.
51. de la Torre, D.; Giacalone, F.; Segura, S.; Martin, N.; Guldi, D. *Chem. Eur. J.* **2005**, *11*, 1267.

Indexes

Author Index

Subject Index

A